ON MONSTERS AND MARVELS

Ambroise Paré at seventy-two

AMBROISE PARÉ

ON MONSTERS AND MARVELS

Translated with an Introduction and Notes by

JANIS L. PALLISTER

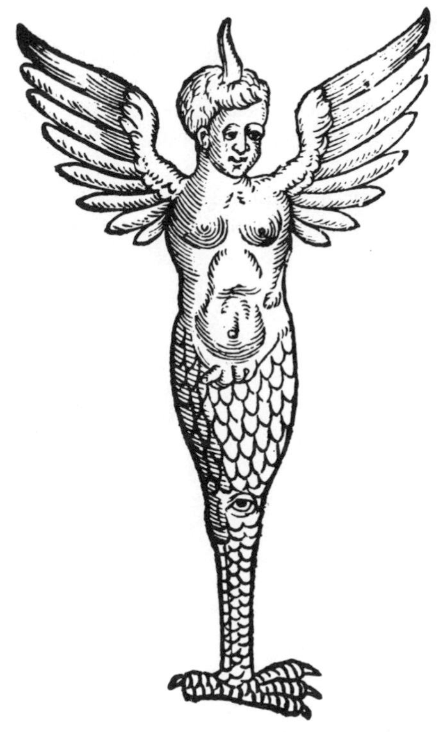

THE UNIVERSITY OF CHICAGO PRESS
CHICAGO AND LONDON

The preparation of this volume was made possible in part by a grant from the Translations Program for the National Endowment for the Humanities, an independent federal agency.

THE UNIVERSITY OF CHICAGO PRESS, CHICAGO 60637
THE UNIVERSITY OF CHICAGO PRESS, LTD., LONDON

Printed in the United States of America
5 4 3 2 1 82 83 84 85 86 87 88 89

JANIS L. PALLISTER is University Professor of Romance Languages at Bowling Green State University. She is the author of numerous publications, including monographs, translations, and four books of poetry.

This translation of Ambroise Paré, *Des Monstres et prodiges,* is based on the Malgaigne edition of 1840.

LIBRARY OF CONGRESS CATALOGING IN PUBLICATION DATA

Paré, Ambroise, 1510?–1590
On monsters and marvels.

Translation of: Des monstres et prodiges.
Bibliography: p.
Includes index.
1. Abnormalities, Human—Early works to 1800.
2. Abnormalities (Animals)—Early works to 1800.
I. Pallister, Janis L. II. Title.
QM691.P3713 599.02′2 81-16297
ISBN 0-226-64562-2 AACR2

For Barb and Willie

Marvels, not Monsters

CONTENTS

List of Illustrations — ix
Acknowledgments — xiii
Introduction by Janis L. Pallister — xv
On Monsters and Marvels — 1
Preface — 3
1. The Causes of Monsters — 3
2. An Example of the Glory of God — 4
3. An Example of the Wrath of God — 5
4. An Example of Too Great a Quantity of Seed — 8
5. On Women Who Carry Several Children during One Pregnancy — 23
6. On Hermaphrodites or Androgynes — 26
7. Memorable Stories about Women Who Have Degenerated into Men — 31
8. An Example of Lack in the Quantity of Seed — 33
9. An Example of Monsters That Are Created through the Imagination — 38
10. An Example of the Narrowness or the Smallness of the Womb — 42
11. An Example of Monsters That Are Formed, the Mother Having Remained Seated Too Long, Having Had Her Legs Crossed, or Having Bound Her Belly Too Tight While She Was Pregnant — 43
12. An Example of Monsters Who Are Created, the Mother Having Received Some Blow or Fall, Being Great with Child — 46
13. An Example of Monsters That Are Created by Hereditary Diseases — 46
14. An Example of Monstrous Things Which Have Occurred in Accidental Illnesses — 48
15. Of Stones That Are Engendered in the Human Body — 49
16. On Certain Monstrous Animals That Are Born Abnormally in the Bodies of Men, Women, and Small Children — 53
17. On Certain Strange Things That Nature Repels through Her Infinite Providence — 60
18. An Example of Several Other Strange Things — 64

19. An Example of Monsters Created through Corruption and Putrefaction — 66
20. An Example of the Mixture or Mingling of Seed — 67
21. An Example of the Artifice of Wicked Spital Beggars — 74
22. The Imposture of a Woman Beggar Who Pretended to Have a Canker on Her Breast — 74
23. The Imposture of a Certain Beggar Who Was Counterfeiting a Leper — 76
24. About a Hedge-Whore Beggar-Woman Pretending to Be Sick with Saint Fiacre's Disease, and a Long Thick Gut Made by Trickery Came out of Her Bum — 81
25. About a Fat Wench from Normandy, Who Pretended to Have a Snake in Her Belly — 83
26. An Example of Monstrous Things Done by Demons and Sorcerers — 85
27. About Those Who Are Possessed of Demons, Who Speak in Various Parts of Their Bodies — 88
28. How Demons Inhabit Quarries or Mines — 89
29. How Demons Can Deceive Us — 91
30. An Example of Several Diabolical Illusions — 94
31. On the Art of Magic — 95
32. On Certain Strange Illnesses — 97
33. On Incubi and Succubi According to Physicians — 105
34. On "Point-Knotters" — 105
Other Stories Not off the Subject — 106
35. Concerning Marine Monsters — 107
36. On Flying Monsters — 136
37. On Terrestrial Monsters — 141
38. On Celestial Monsters — 150
39. [On Natural Disasters] — 158
Appendix 1. Items from the *Discourse on the Unicorn* — 163
Appendix 2. From *The Book of Tumors* — 170
Appendix 3. From *The Sicke Womans Private Looking-Glasse* — 174
Suggested Identifications — 177
Notes — 181
Bibliography — 208
Index — 213

ILLUSTRATIONS

Ambroise Paré at 72
Frontispiece
A Fifteenth-Century Anatomy Lesson
page xiv

1. Figure of a colt with a man's face — 6
2. Figure of a winged monster — 7
3. Figure of a girl having two heads — 9
4. Figure of two twin girls joined and united by the posterior parts — 10
5. Figure of a man from whose belly another man issued — 11
6. Figure of the horned or hooded monster — 12
7. The shape of a monster found in an egg — 13
8. Figure of a child having two heads, two arms, and four legs — 14
9. Figure of two twins having only one head — 15
10. Figure of two twin girls joined together by their foreheads — 16
11. Figure of two monstrous children born not long ago in Paris — 17
12. Figure of two girls joined together by their anterior parts — 18
13. Figure of a monster having two heads, one male and the other female — 19
14. Figure of a monster with four legs and four arms — 20
15. Figure of a man having a head in the middle of his belly — 21
16. Figure of two very monstrous infants, in whom only one set of female sexual organs is manifested — 22
17. Figure of a monstrous pig, born in Metz in Lorraine — 22
18. The picture of Dorothy, pregnant with many children — 25
19. Picture of a hermaphrodite man-and-woman — 28
20. Figure of two hermaphroditic twin children, being joined back to back the one to the other — 29
21. Figure of a monster having four arms and four feet and two female "natures" — 30
22. Figure of a monster having two heads, two legs, and only one arm — 34
23. Figure of a monstrous child, coming from a lack of seed in proper quantity — 35
24. Figure of a female monster without a head, front and back views — 36
25. Figure of a man without arms — 37
26. Two figures, one of a furry girl, and the other of a child that was black because of the imagination of their parents — 39

27. Figure of a very hideous monster having the hands and feet of an ox, and other very monstrous things — 40
28. Prodigious figure of a child having the face of a frog — 41
29. Figure of a child who has been pressed against the mother's belly, having his hands and feet twisted — 44
30. [Figure of dead fetus carried in mother's womb for twenty-eight years] — 45
31. Figure of a stone extracted from a pastry cook of Montargis — 50
32. Figures of three stones extracted at the same time without any time interval, from the bladder of a man called Tire-vit, one of which is broken — 51
33. [Figure of an animal in an apostema (ed. Malgaigne)] — 56
34. [Figure of an animallike material ejected by Count Charles de Mansfeld (ed. Malgaigne)] — 56
35. Figure of a child who had a live snake in its back — 58
36. Figure of a piece of mirror, which a child twenty-two months of age swallowed, which was the cause of his death — 62
37. Figure of a child, part dog — 68
38. Figure of a monster with the face of a man and the body of a goat — 69
39. Figure of a pig, having the head, hands, and feet of a man and the rest of a pig — 70
40. A monster half-man, half-swine — 70
41. Figure of a monster like a dog, with a head like a bird — 71
42. The figure of a lamb having three heads — 72
43. Picture of a Triton and a Siren, seen on the Nile — 108
44. Marine monster having the head of a Monk, armed and covered with fish scales — 109
45. Figure of a marine monster resembling a Bishop dressed in his pontifical garments — 110
46. Figure of a marine monster having the head of a Bear and the arms of a Monkey — 111
47. Figure of a Marine Lion covered with scales — 111
48. Picture of Marine monster having a human torso — 112
49. Hideous figure of a Sea Devil — 113
50. Figure of a Sea Horse — 113
51. Figure of a Marine Calf — 114
52. Figure of a Marine Sow — 115
53. Figure of a fish named Orobon — 116
54. Catching Crocodiles — 117
55. Two fish, one like a plume and the other like a bunch of grapes — 118

56. The "Aloés," a monstrous fish 119
57. Snail from the Sarmatian Sea 120
58. Portrait of the Hoga, a monstrous fish 122
59. Portrait of certain flying fishes 123
60. Figure of another very monstrous flying fish 124
61. Diverse shells, together with the fish which is within them, called Bernard the Hermit 126–27
62. Figure of the Lamie 128
63. Figure of the fish called Nauticus, or Nautilus 129
64. Figure of a Sea Crab 130
65. Figure of a whale-catch 132
66 Another species of Whale [1573; 1579] 133
67. Figure of an Ostrich 137
68. Portrait of an Ostrich skeleton 138
69. Figure of the bird named Toucan 140
70. Figure of the Bird of Paradise 141
71. Figure of an animal called Huspalim 142
72. Figure of the Giraffe 143
73. Figure of the Elephant 144
74. Figure of the animal Thanacth 146
75. Figure of a monstrous animal which lives only on air, called Haiit 147
76. Figure of a very monstrous animal that is born in Africa 148
77. Figure of the Chameleon 149
78. The figure of a fearful comet seen in the air 151
79. Figure of a Comet 153
80. Figure of the Marine Boar 164
81. Figure of a Sea Elephant 164
82. Figure of the Pyrassoupi 165
83. Figure of the Camphurch 166
84. Figure of the Florida Bull 167
85. Figure of the Rhinoceros 168
86. Alternate Figures of the Ravenna monster 185
87. The double monster of Rhodiginus and that of Lycosthenes 186
88. Homo signorum 197
89. A. Dürer's Rhinoceros 207

ACKNOWLEDGMENTS

This book was completed with the assistance of a grant from the National Endowment for the Humanities, a sabbatical from Bowling Green State University, and a fellowship from the Bowling Green State University Faculty Research Committee. I am indebted to the Department of Romance Languages of Bowling Green State University for financial assistance in the preparation of this book and the libraries of Bowling Green State University for their diligent attention to my research needs. I should especially like to acknowledge here the help of the librarians Kusalya Padmarajan, Kay Sandy, Angela Poulos, Dawn McCaghy, and Dwight Burlingame. In addition, I thank H. Douglas Lamb, Fellow of the Royal College of Physicians of Canada, for helping to establish the bibliography. Recognition is also to be extended to Librairie Droz, Geneva, for permission to reproduce the figures from Jean Céard's critical edition of Paré's *Des Monstres*.

For expert opinions on different problems regarding the book I should like to credit Anna Miller, Micheline Ghibaudo, Michael Giordano, and Armand A. Renaud, for their advice in matters of French, and Richard Hebein, professor of classics, and Ralph Frank, professor of geography.

For additional assistance and for support of this project I should like to thank Professors Jürgen Herrmann, Ramona Cormier, Henri Peyre, David Lee Rubin, Armand A. Renaud, Leonard Wilson, and Pierce Mullen.

A fifteenth-century anatomy lesson, from Mondino's Anathomia *(Venice, 1493). While the professor reads from the text, his assistant points out the organs under discussion.*

INTRODUCTION

In epitomizing the contributions of Montaigne to the world of thought, Sainte-Beuve observed that there is something of the great essayist in each one of us. Surely, very much the same statement could be made of Ambroise Paré, as he emerges in his book *Des Monstres et prodiges*. Far from being simply a technical treatise on birth defects, the work is a scientific and literary document that reveals the scope of the man and, moreover, manages to be, as Malgaigne claimed, "one of the most curious books of the French Renaissance."

In *Des Monstres* we can see many of the modern sciences in their embryonic form. Surgeon above all, Paré nonetheless shows himself in this particular work to be interested in what today would be the province of many other branches of medicine and science. Moreover, in the more "humanistic" categories that interested Sainte-Beuve, one might consider Paré as a philosopher, as a historian and storyteller, as a critic, as a philologist, as a moralist (especially concerned, in this book, with the malingerer), as a demonologist who displays a fundamentally Christian attitude, and so forth.

One can, however, find a synthesizing thread in all these seemingly disparate trends. Paré's essential focus is on the architecture—the infinitely varied shapes and sizes—of nature, its engineering, especially as regards mechanics and anatomy, its order (the normal), and its accidental anomalies and disorders, including unusual size and proportion, irregularity, and rare or strange phenomena, in short, all its "monsters and marvels." Similarly he is concerned with supernatural occurrences—in which, when "brought about through witches," Paré did not have so much faith as Lyons and
Petrucelli suggest. 1

A fundamentally theocentric interest in a multiplicity of subjects, especially in astronomy and philosophy, as demonstrated and advocated by Paré, is, however, a common characteristic of Renaissance surgeons. (See, for example, Johannes de Vigo or Hieronymous Von Braunschweig.) Paré's frequent referral of the mysteries of disease and healing to an omnipotent deity—summed up in his celebrated statement *Je l'ay pensé, Dieu l'a guari,* or "I cared for

2 him; God cured him"—is common among Renaissance surgeons
who were, for the most part, a fairly devout group, whether Prot-
3 estant (as Paré may have been) or Catholic. But more importantly,
the proportions of Paré's four-square, four-part book, evoking the four seasons and the four cardinal points of the world, his use of illustrations, and his attention to detail, especially to descriptions of dimension and structure within the narrative passages, are characteristic of the great physician or surgeon, the great scientist, and the great teacher in almost all ages and cultures. Even his particular applications of the theories of the microcosm and the macrocosm reflect his fascination with conformation and structure.

Despite the repeated accusations of frequent "borrowings" from other writers, Paré's composition and style, as well as his unifying concern with pattern and structure, make *Des Monstres* not just another of the many treatises on "monsters" written during the period, but rather *the* book on "monsters," for it, perhaps better than any other, presents a synthesis of views and theories on this and other subjects, while at the same time illuminating one of the richest, albeit rough-hewn, minds of sixteenth-century Europe.

Ambroise Paré was born in Bourg-Hersant near Laval in Maine, probably in 1510. Though little is known of his family, he had at least one sister, Catherine, who married Gaspard Martin, a barber-surgeon in Paris, and two brothers, both named Jehan, one a chest-maker in Paris and the other a barber-surgeon who practiced in Vitré in Brittany.

We know very little about Paré's education. By his own admission he never acquired Greek or Latin, having learned his Galen through a "French interpretation" by Jehan Canape, physician on the faculty at Lyons, who also "interpreted" other masters of the art of medicine. By the age of fifteen (in 1525) Paré was in Angers, where he may have been a student. But there is no certainty about where or with whom he may have studied surgery. Perhaps with Vialot or Laval? Or with his brother in Vitré? Or in Angers? Or even in Paris? What *is* known, however, is that his life as an apprentice surgeon cannot have been pleasant, for accounts of the period show that apprentice surgeons were cruelly treated and that their condition was one of virtual slavery. They were obliged to do all the shaving of perhaps fifty customers each day beginning early in the morning and to work on their studies between all the tasks such a busy schedule commanded. They would therefore rise as early as four o'clock in the morning to attend university lectures given in Latin, which they did not understand.

By 1533, at about twenty-three years of age, Paré was in Paris. He became *compagnon chirurgien* or *aide-chirurgien* at the Hôtel-Dieu, the chief public hospital of Renaissance Paris, and, as house surgeon, he lived for three or four years inside this hospital. Sylvius (or Dubois) was one of his teachers.

After 1537, Paré divided his life for thirty years between serving in the army during times of war (described in his *Journeys in Diverse Places*) and caring for the sick of Paris during the intervals of 4
peace. By 1541, when he was about thirty-one, he had qualified as a master barber-surgeon. We think of him as drawing heavily upon observation for his knowledge, a practice that is discernible even in *Des Monstres*—where he quotes others more than he may be inclined to do in his more strictly surgical and medical treatises. In any case, direct observation is a salient feature of his approach—one, along with quick judgment, that led him to realize that boiling oil was not the correct treatment for gunshot wounds and one he 5
himself emphasized in his great *Apologia,* a literary masterpiece in which he says to Gourmelen, his detractor, "Now will you dare to say you will teach me to perform the works of Surgery, you who never yet came out of your study?" 6

On 30 June 1541, Paré married Jehanne Mazelin in his parish church of Saint André des Arts in Paris. He had three children by her: a son François, who was baptized on 4 July 1545 at Saint André and died a few months later; a son Isaac, baptized at Saint André on 11 August 1559, who died and was buried 2 August 1560; and a daughter Catherine, baptized at Saint André on 30 September 1560, who lived to marry and give him grandchildren. On 4 November 1573, his first wife, Jehanne, died; she was buried the same day at Saint André.

The strife among physicians, surgeons and barber-surgeons of
7 the period was endemic, and although Paré avoided the national politics of his time fairly well, he was unable to escape professional conflicts. By 1554 he had been advanced from barber-surgeon to surgeon; for in that year on 23 August he had been admitted into the confraternity of Saint-Côme, the Royal College of Surgeons, and given the degree of bachelor. The confraternity, being more than glad to have him in their company, had made of the examinations, normally conducted in Latin, something of a formality; moreover, on 8 October he had been examined for and granted the degree of licentiate. On 17 or 18 December the degree of master was conferred upon him. As Paget observes, the procedures in admitting Paré to the confraternity were what even Aeschylus
8 long ago had called a "plant," and the faculty of medicine of the university was well aware of it. But his battles with the faculty had only just begun, and we shall return to them shortly.

Paré was surely a man of business. He bought up the property around his own home in the parish of Saint André des Arts in Paris, until he owned a number of adjoining houses just off the Seine between the rue (now quai) des Augustins and the rue de l'Hirondelle. We know that he also owned a house and vineyards in Meudon on the rue des Pierres, a fact he alludes to in chapter 19 of *Des Monstres*. In addition he owned a house on the rue Garancière in the Faubourg Saint-Germain and a property at Ville-du-Bois outside Paris. Furthermore, records still extant show his shrewdness in fixing dowries on his daughter Catherine, his niece Jehanne, and his second wife Jacqueline and in caring for his own. It is apparent that his family was a very close one.

Paré lived in very troubled times and he saw the vicious civil wars through from beginning to end. Having become chief surgeon to Charles IX on 1 January 1562, he survived the Saint Bartholomew's Day Massacre of 22 August 1572 through the protection of Gaspard de Coligny, admiral of France, and that of the king. Where he stood in these religious controversies is not certain. That he was not a devout Catholic comes through in some of the stories told of him; that he was a humble man and a believer, however,

is everywhere clear in his writings. Was he secretly in sympathy
with the Huguenots? The Haags, in their treatment of him in the
celebrated *France protestante,* seek to demonstrate that this was the
case. D'Eschevannes's and Paget's arguments for his fundamental 9
Catholicism seem more convincing. 10

In any event, overriding the obstacles offered by his busy life
and by the turmoil of the times, Paré, within a matter of a few
months following his first wife's death, was able to find another.
The haste of the marriage suggests a certain pragmatism, for Paré
had, after all, been left alone with the care of his daughter Catherine
and his niece Jehanne. It is curious to note that on 18 January
1574, Paré married Jacqueline Rousselet, the daughter of Claude
Rousselet, dean of the faculty of medicine in 1577, with what Paget
calls "more money and less romance" than obtained in his first
marriage, and in 1581 his daughter Catherine married Jacqueline's 11
brother François Rousselet, by whom she had eight children. Catherine died 21 September 1616.

By Jacqueline, Ambroise Paré had six children: Anne was baptized at Saint André 16 July 1575 and never left progeny; Ambroise (I) was baptized 30 May 1576 and died in infancy; Marie was baptized 6 February 1578; Jacqueline was baptized 8 October 1579 and died in 1582; Catherine (II) was baptized 12 February 1581 and died in 1659, having given birth to twelve children, one of whom was the famous François, Abbé d'Aubignac; and finally Ambroise (II), who was baptized on 8 November 1583 and died August 1584. One must observe that Paré never had a son live to adulthood, to whom he might have entrusted his skills. A nephew, Bertrand, whom he had taken under his wing, turned out to be something of a ne'er-do-well.

Despite Paré's status as chief surgeon to Charles IX, who died
of phthisis on 30 May 1574, and to Henri III, under whom Paré
also became a member of the king's council, the years 1575–85
were years of great strife for him, for he was here waging what
Paget called a "sort of Holy War for the deliverance of surgery
from the bondage of medicine." The battle was waged over the 12

publication of his *Collected Works,* which appeared in three editions, in 1575, 1578, and 1582. (The 1582 edition was a Latin edition done by Jacques Guillemeau.) The faculty of medicine, headed by Estienne Gourmelen, was outraged by the 1575 edition, chiefly because the insolent Paré had dared to write such a work, that is, to deal with questions beyond his competence, not being a physician—even though they had never objected to his publishing his treatises on surgery as separate entities prior to this. Furthermore, they claimed that his book contained much that was grossly indecent and immoral. They objected to the use of French and to the illustrations on the grounds that women and children could read the work and see the pictures, and this even though Boaistuau and others had published most of the illustrations accompanied by texts in French before: Paré had only borrowed them. By 1577 the persecutions of the faculty had lessened, perhaps in part because Paré's father-in-law was then dean. But the Latin edition did further infuriate them, apparently only because someone other than they had dared to translate a book into Latin. (Guillemeau was, like Paré, a surgeon.) Hamby holds that the faculty was convinced that Guillemeau had not done the translation, but rather a physician of the faculty, who had gone unrecognized (Jean Haultin?). The fracas was indeed a silly one, but it illustrates the kind of rivalry that existed among the various branches of the healing arts during Paré's time and indicates the kind of stress Paré had to deal with throughout a large segment of his life. Interestingly enough, the great, authoritative 1585 edition of his work brought little opposition from the faculty. It was the last edition published during his lifetime.

Ambroise Paré died in December 1590 (at the age of eighty) only a few months after his confrontation with the Archbishop of Lyons on behalf of the starving people of Paris. His death is eloquently described by Pierre de L'Estoile, who makes the following entry in his memoires:

> Thursday, December 20, 1590, Ambrose [*sic*] Paŕe, surgeon to the King, died in his house in Paris, at the

> age of eighty. He was a learned man, foremost in his art, who, despite the times, spoke freely for peace and for the public welfare, which made him as much loved by good men as hated and feared by the wicked—which latter far outnumber the others, especially in Paris, where the rebels have all the authority. 13

Ambroise Paré was buried, at his own request, at the foot of the nave, near the bell tower of the church of Saint André des Arts, which was demolished in 1807. His remains were transferred without special consideration to a common grave. Paré's second wife, Jacqueline, survived him, and lived until 26 June 1606.

Although he wrote as long ago as 1899, Paget, in assessing Paré's life and work, evokes the mettle of the man when he writes, "The breadth, insight, force and humanity of his writings, their shrewd humour, his infinite care for trifles, the gentleness and clear-headed sense of his methods, they are amazing. It is no answer to say that Paré was ignorant, superstitious, credulous, bound hand and foot by medieval imagination and tradition. Truly his explanations are childish, and his ignorance of things not yet discovered is as profound as our own: but put Ambroise on the side of the patient's bed, and a surgeon of our own day, singlehanded, on the other: you will not find the balance of insight and practicality against Ambroise." In view of the strides made since Paget's time 14
in the art of surgery, some critics might not be willing to accept such a sweeping endorsement of Paré's abilities. Others will agree with it.

Paré's contribution was in any case considerable. In addition to his two great discoveries regarding, first, the negative effect of using boiling oil in gunshot wounds, and second, the use of ligation to replace cauterization in amputations, he is a veritable mine of modern precepts. A "humanist" in the broadest sense of the word (unlike such contemporary medical humanists as Nicolao Leoniceno, Johann Guinter of Andernach, or Thomas Linacre, who were engaged in the recovery of the ancients), Paré was concerned with 15
the suffering of all mankind regardless of creed or politics. His

treatise on the plague reaches great poetic heights in its empathetic description of the straits in which the victims of this dread disease might find themselves; it can be sampled in English from the stirring translation of Stephen Paget. Paré is the author of a work on embalming corpses, which is the earliest formal treatise on medical jurisprudence. In addition, Paré extolled sanitation and cremation in cases of the plague or other infection. Moreover, he suspected that infection was air-borne; he used damp antiseptic dressings, evolved a local anaesthesia, opposed immoderate bleeding, and in particular stressd the value of silence for the ill.

Paré was not a narrow humanist but an empiricist and a practical man who read "little Latin and no Greek" but who insisted upon experience/experiment as the mistress of arts and sciences, a precept ironically derived from Aristotle. This, above all, made him a modern in the great Quarrel of the Ancients and the Moderns, which reigned throughout the sixteenth and seventeenth centuries in France. Indeed, in 1545 Paré used the formula, "the Ancients must serve as our beacons to see further into the distance." We can discern this interest in experiment, embryonic though it may be, in *Des Monstres;* note, for example, his tendency to collect specimens and to pose questions, evidenced in many passages. But perhaps one of the most striking facets of Paré's doctrine is the stress he gives to moderation, which, in this respect, not only makes
16 him, like Montaigne, a forerunner of classicism, but which also brands him as a true Frenchman and reveals him to be a person of measure, interested in, but also offended by, "monstrous" departures from the mean, especially if they could be avoided through self-discipline.

The long history of the depiction of monsters and deformities throughout antiquity and the Middle Ages has been ably traced by Jean Céard in *La Nature et les prodiges* and more especially by Jurgis Baltrušaitis in his remarkable study entitled *Réveils et prodiges*. As I have already noted, by the end of the sixteenth century treatises on monsters had become a veritable genre. Therefore, what concerns us in particular here is Paré's contribution to that genre and the evolution of his book about monsters and marvels within his own works. Baltrušaitis seems to have placed Paré in the history of teratology best when he wrote, "The fantastic realism of the image-makers is joined here to the awakening of a realistic [or scientific] mind." 17

Taking cues, then, from some of the books on natural history and on monsters that he was familiar with, Paré began to be interested toward 1570 in this question, as one connected with human reproduction. It was through a conversation with and at the request of the Duke d'Uzès, a peer of France, that Paré first undertook to write his 519-page work on reproduction, which was published in 1573 by André Wechel of Paris. It was divided into two parts:

> *Deux Livres de chirurgie.* I. *De la generation de l'homme, et manière d'extraire les enfans hors de la mère, ensemble ce qu'il faut faire pour la mieux faire et plus-tost acoucher, avec la cure de plusieurs maladies qui luy peuvent survenir.* II. *Des monstres tant terrestres que marins avec leurs portraits. Plus un petit traité des plaies faites aux parties nerveuses.*

The very title shows the primitive state of the treatise on monsters at this juncture. The celestial monsters are not yet present, nor is the ultimate four-part nature of the work apparent. Yet the same title shows that from the beginning Paré tied monsters closely to human reproduction and that for him consideration of monsters was from the outset a matter of "scientific" inquiry.

Following this primitive edition under André Wechel, the *Oeuvres de M. Ambroise Paré, conseiller et premier chirurgien du roy, avec les figures et portraicts tant de l'anatomie que des instrumens de chirurgie et de plusieurs monstres* appeared in 1575 (Paris: Gabriel Buon). New editions of this work appeared in 1579 and 1585. The latter contains the *Apologia,* in which Paré responds to his critics, especially to Gourmelen. Malgaigne gives us an account of that dispute, which had begun with the appearance of the 1573 edition.

18 According to Malgaigne, Gourmelen had been elected dean of the faculty of medicine in 1574. Wounded by the fact that Paré's *Cinq livres de chirurgie* (1572) had overridden his own *Synopsis chirurgiae,* published in 1566 and translated into French in 1572, he sought redress by invoking an old law of 1537 which forbade the publication of any book of medicine that had not received the approval of the faculty. He gathered the faculty around him and sought to block the sale of Paré's book—already in print. The faculty did not complain about the crude expressions found in Paré's work. That was to be expected, given the unformed state of the technical language, coupled with the tenor of the times. The faculty also admitted Paré's right to discourse on surgical matters. What it condemned was that he had tackled the great questions of philosophy and medicine (for example, the elements, the humors, the faculties, and so forth), and that in his book he had also "discoursed on sperm, menstrual blood, causes and signs of conceptions, and the like"—all these being essentially medical matters, which the faculty regarded as its property. Particularly galling to
19 the faculty was that the work contained a book on *fevers!* Although some legal steps were taken, little really came of the faculty's efforts; the book remained essentially unchanged and went on sale.

One should observe that the faculty had other objections that could not be made a part of its legal case, such as Paré's use of French rather than Latin, since use of the mother tongue would reveal medical secrets to everyone (especially to women and girls). But these physicians were even more perturbed by the fact that Paré made the humiliating claim in his work that surgery, by its antiquity, need, certitude, and difficulty surpassed internal medicine. Even some of Paré's fellow surgeons reproached him for having made surgery available to anyone and everyone who could read. Yet all of this is most ironic, for Paré's book was the most complete work on surgery ever to have been written. According to Malgaigne, even that of Guy de Chauliac, more erudite and methodical, was not so original or so thorough. Indeed, Packard finds that Paré's written work marks an epoch, for only a very meager surgical literature existed prior to his. Moreover, Packard contends that

Paré is a true Renaissance man who sought synthesis, introducing clear innovation into the healing arts with his effort to bring medicine and surgery into their proper relationship to one another, to underscore the need surgeons had of medical training, and to demonstrate the scientific nature of surgery. Packard might have added that all of this was doubtless due to an underlying Renaissance trait that spurred Paré on: the spirit of critical investigation.

In 1582 Paré published a new work in which he doubted the efficacy of the use of unicorn and mummy in the treatment of illnesses and injuries, and in 1583 a response was published which had been approved by Grangier, dean of the schools of medicine. The anonymous author derided Paré, saying, "I do not know what to do in place of all those monsters you have inserted into your surgery, which are completely off the subject and which are fit for amusing little children, except to paint you here going on all fours." In his *Apologia,* which was first published separately, and then again in the 1585 edition of his *Oeuvres complètes,* Paré responded with measure. Paré defended himself with calmness and self-assurance, here using his detractors' own arguments against them:

> For my part I esteem nothing in my book pernicious because it is written in our vulgar tongue. Thus the divine Hippocrates wrote in his language, which was known and understood by women and girls, talking no other language than that. As to me, I have not written except to teach the young surgeon, and not to the end that my book should be handled by idiots and mechanics, even if it was written in French. [Le Paulmier, pp. 222–48]

"To teach the young surgeon"! Though some will see in Paré's book on monsters a pedantry that goes far beyond pedagogical needs, the didactic purpose will stand out for others. In any case, it is useful to keep this intention uppermost in our minds when we approach the book *On Monsters.* It is true that Paré used many sources, including Lycosthenes, Boaistuau, Thevet, Pliny, Aristotle, Claus Magnus, Du Bartas, etc. All this borrowing caused him to

be accused of something approaching plagiarism, and yet Paré's book on monsters had a new element, overlooked at the time. For, as Jean Céard points out, despite Paré's rapid treatment of the *end* or *purpose* of the existence of the monstrous (e.g., "signs of some
20 misfortune to come"), that is not at all where his true interest lies. Nor is he concerned with rigorous definitions of "monster," "marvel," or "the maimed." Rather, his basic concern is his search for causes, for why the normal form has failed to occur. This is the true design and purpose of his book, which, appearing at first blush to be disorderly, in reality has a careful plan, and even a method, one that escaped Malgaigne. For though without an experimental approach, Paré is intent upon accumulating data regarding these causes. Moreover, Céard has quite correctly pointed out that in *Des Monstres* the causes are discussed by analogies, and, more importantly, are arranged simultaneously, from high to low and
21 from the exterior to the interior. We might reiterate, too, that *Des Monstres* has a clear division into four parts: (1) physical and moral monstrosities in the human and animal domain, (2) flying monsters or oddities, (3) terrestrial monsters or oddities, and (4) celestial monsters. Reference to earth, water, air and fire—the elements then known to man—clearly comprises one of the book's infrastructures.

Moreover, Céard is again right in concluding that, when dealing with fishes, birds, and animals, Paré's concerns are not the taxonomical ones of his predecessors. Instead, like Thevet, Paré dismisses any widely known creature and concentrates on the rare. "The 'aloés,' more than the Pike or the Salmon, gives him the
22 feeling of the incomprehensible power of Nature and of God." Similarly, Paré gives little space to the theory that celestial phenomena are presages; he is not concerned with their meaning or end but with their very existence. And, one might add, they provide for him the proof, when taken together with all the other rare and unusual shapes and forms he had discussed throughout the book, of the infinte variety of God's universe. All forms—icthyomorph, zoomorph, teratomorph—celebrate the endless variation and di-

versity which God, in his creativity and infinite power, has allowed to flourish and multiply.

Thus, as Paré's book on monsters evolved over a decade or so, the changes that occurred were dictated by these major themes and premises: a concern for causes (not ends), a fascination with the rare, and an appreciation for the varied—all and always promulgated with an almost classical intention to instruct and to inform.

Again, it is Céard who has most clearly located the impetus and
most correctly assessed the contribution of Paré's work: "the book
Des Monstres et prodiges of Ambroise Paré can be considered the
most sustained attempt (during the sixteenth century) to 'naturalize'
monsters." And finally, it is this effort to naturalize, joined to the 23
idea of variety, that permits "the scandal associated with monsters
to come to a halt, to such a point, in fact, that monsters even
become the most notable signs" of variety. 24

While there have been several French editions of *Des Monstres* since it was first published in 1573, and especially the great critical edition of Jean Céard (1971), that of Malgaigne (1840) constitutes the basic text used in this translation. In preparing his critical edition of the *Traité des Monstres,* which was to be the nineteenth book of Paré's *Surgery* (or *Oeuvres complètes*), Malgaigne followed the 1598 (posthumous) edition of Paré's *Oeuvres complètes,* but he also referred to the 1585 edition. Occasionally, Malgaigne shows variants from the very early 1573 and 1579 editions of *Des Monstres,* and for the most part, these variants have been incorporated into the present text in parentheses. Similarly, Paré's original footnotes have been incorporated into the text in parentheses.

In the 1582 Latin edition of Paré's *Oeuvres complètes,* parts of *Des Monstres* were moved from their original setting to the *Discourse on the Unicorn.* These variations, retained but pointed out by Malgaigne, have been included in the present translation as Appendix 1.

It is strange that Paré's *Des Monstres* has never before been translated from the original French into English. Johnson's early and very crude English translation of Paré's writings was made from the 1582 Latin version of Paré's *Works.* And about the Latin version a word is in order here. Made by Jacques Guillemeau, a surgeon and former student of Paré, it contains many very significant variants and lacunae, but, more important, of course, it lacks the additions and omissions of the 1585 French edition. Moreover, the Latin version omits anything that posed problems in comprehension, which can occur rather frequently, for Paré's style can be elliptical and even incoherent at times.

Now, from this extremely faulty Latin translation Thomas Johnson made his English translation of Paré's works. These included, of course, the deformed treatise on monsters, and they were published under the title of *The Workes of That Famous Chirurgion Ambrose Parey, Translated out of Latine and Compared with the French* (London: Th. Cotes and R. Young, 1634). But it is obvious that the English translation of a translation of the incomplete edition of *Des Monstres*—in no way critical and often inaccurate and unfaithful (see Doe)—can serve the scientist or the humanist no very great purpose in finding out Paré's true thoughts on the monstrous in nature. As we come to recognize that Paré is a central figure in the study of French Renaissance civilization, careful translation and editing of Paré's writings become more and more essential.

Though *Des Monstres,* then, has remained untranslated until now, one might call attention to two modern translations of Paré's *Apologie* and his *Voyages faits en divers lieux,* which give us a good deal of information about Paré, both in the military and in civilian life. They are Paget's *Journeys in Diverse Places,* in *Ambroise Paré and His Times* (also published in *Scientific Papers*), and Packard's *Life and Times of Ambroise Paré,* which contains a translation of *The Apology and Treatise* (containing *The Voyages Made into Divers Places*). In addition, many passages from Paré's treatise *On the Plague* can be found in translation in Paget's *Ambroise Paré and His*
25 *Times.*

The illustrations for this edition (except figs. 33 and 34) have been taken from Jean Céard's critical edition of *Des Monstres* and are reproduced with the kind permission of Librairie Droz, Geneva. It is to be noted that few of these illustrations were ordered or

prepared by Paré himself (one exception being the diagram of the ostrich skeleton). Most are taken from various studies that Paré consulted as he was writing *Des Monstres*. These included figures borrowed chiefly from Lycosthenes (who often borrowed in his turn from Rueff, Boaistuau, Benivenius, Gesner, etc.; see notes). Although the illustrations were regarded by Malgaigne as a blemish (*une souillure*), they constitute an important aspect of Paré's interest in form and suggest an emblematic preoccupation characteristic
of the age, as well as a ramification of Paré's pedagogical and 26
dogmatic intentions, for he sought everywhere to teach the young surgeon, who was to be well informed in *all* subjects. The figures presented here are, then, suggestive of Paré's motives and interests as they evolved over a decade or more. It seems unfortunate, therefore, that Malgaigne can have suppressed so many of the illustrations, deeming them "absurd," "ridiculous," "inapplicable to surgical concerns," and so on; no edition of *Des Monstres* can be considered complete and authentic without the full complement of figures. That is another reason why the Johnson translation from the Latin must be rejected; for not only are the figures incomplete, but they are often reversed, and some appear to have come from sources other than those selected by Paré. In any event, they are almost invariably not the ones that accompany the 1585 edition.

One of the most striking things about Paré's language, and one that has won him a place in most discussions of *la langue verte* in the French Renaissance, is his frank and forthright use of words that might be considered quite vulgar. Oddly enough, he intermingles these freely with technical terms. Hence, on the same page or within a few pages we find him using, almost indifferently, the words *pisser* and *uriner,* or *fleurs* and *mois* juxtaposed to *menstrues.* In referring to the penis he shows equal vacillation, preferring the word *verge,* which has here been translated as *rod,* despite the more favored Renaissance English translation, *yard.* Such vacillation between the vulgar and the technical terminology is not so surprising, however, if we remember that Paré was not a thoroughly educated man and, more importantly, that he was writing for even less

educated barbers and surgeons. Moreover, the technical medical language was still unfixed at this period, though there is evidence
27 of an ever growing lexicon for this specialized purpose.

Aside from Paré's occasional use of coarse language there is nothing particularly noticeable about it; yet his style is sometimes extremely difficult because of the degree of its incoherence, and, while I have corrected the sentences for easier reading, I have done so in brackets, so that the reader of the English text may appreciate the chaotic nature of Paré's syntax.

28 Of Paré's style Malgaigne has written at length, but I may sum him up by saying that the nineteenth-century editor complained chiefly about Paré's lack of polish and, more particularly, about his lack of solid sentence structure and coherence. Surely the person who undertakes a translation of Paré will be especially conscious of this, but, even so, Paré's style has some beauties and virtues that must be pointed out.

In *Des Monstres,* and also in the *Diverse Journeys* (the latter headed by the *Apologie*), Paré is very clearly garrulous, gossipy, and conversational, and he demonstrates a love of storytelling characteristic of the age. In this he is specifically influenced by Noël du Fail. Paré might, upon occasion, make us think of Rabelais, or so Mal-
29 gaigne contends, and yet there is usually a certain sobriety in Paré's style that makes the comparison largely invalid. Yet, Paré's storytelling nature determines that very incoherence we cannot help noticing; the sentences are very disconnected and at times thoroughly obscure. The free drift of Paré's writing is, in fact, a part of its charm. Any reader of *Des Monstres* will, moreover, note the extent to which Paré's style is idiomatic, imaginative, personal. Upon occasion we find him using metaphor, irony, and even the *style indirect libre,* though this is generally associated with nineteenth-century authors, especially Flaubert.

In addition, Paré could upon occasion rise to great polemical and poetic heights; examples of this are not lacking, as in his majestic description of the movement of the heavenly bodies, or the heavily alliterated passage describing a volcano, found in chapter 39. The exact quotation from the original French seems in order here, for when we are speaking of poetry, translations do not seem adequate:

> L'an 1329, le premier iour de iuillet, ayant fait nouuelle ouuerture, abbatit et ruina par ses flammes et tremblement de terre qui en aduint, plusieurs Eglises et maisons situées à l'entour de ladite montagne: elle fit tarir plusieurs fontaines, ietta dans la mer plusieurs bateaux qui estoient à terre, et au mesme instant se fendit encore en trois endroits de telle impetuosité, qu'elle renuersa et ietta en l'air plusieurs rochers, voire aussi des forests et vallées, jettant et vomissant tel feu par ces quatre conduits infernaux, qu'il decouloit de ladite montagne en bas, comme des ruisseaux bruyans, ruinant et abbattant tout ce qu'il rencontroit ou luy faisoit resistance: tout le pays circonuoisin fut couuert de cendres sortans hors de cesdites gueules ardantes au sommet de la montagne, et beaucoup de gens en furent estouffés: de maniere que lesdites cendres de ceste odeur sulphurée furent transportées du vent (qui souffloit alors du Septentrion) iusques à l'Isle de Maltha, qui est distante de 160. lieuës Italiques de ceste montagne là.

Just such poetry, coupled with a profound sense of the tragic, permeates Paré's treatise *On the Plague,* which is necessary reading for the person who would grasp the full mental stature of the man, not to mention his warm compassion and humanity.

Any reader will notice the degree to which Paré quotes the "authorities," such as Hippocrates, Pliny, the Bible. This was a common trait in the period, even among authors not considered to be learned, and while the heavy use of sources may be a personal trait of his style (like the mixing of vulgar and technical language), it is more accurate to say that the procedure was standard among Paré's contemporaries. (Note, for example, Montaigne's abundant quotations from the authorities.) Thus, Paré's lack of training in Latin, Greek, rhetoric, and the like, and his timidity in the face of the great physicians—he being only a surgeon—are less noteworthy than some scholars contend; moreover, they have less bearing on his use of "authorities" than does this contemporary mania for quoting, which he seems to be imitating, almost as if it were a stylistic necessity. However, it is important to remember that the quotations are often inexact and even attributed incorrectly; this suggests that Paré was less interested in the documentation than in the monstrosity itself. In other words, he was chiefly concerned with analyzing and bringing together for his "students" the things he had directly or indirectly observed about the strange and the rare in proportion, size, quantity, or distance. With this in mind, we should not dwell too much upon the accuracy or inaccuracy of his attributions. As Malgaigne says, the thought is ultimately Paré's own, as is the expression in which he chooses to encapsulate it.

The reading of Paré is, then, despite its difficulties, a rare treat for the student of literature and of philosophy; for the style does not lag behind the mind nor does the mind surpass the style. Therefore, one must endorse Malgaigne's strong assertion that "a true literary genius is unexpectedly revealed in him."

Effort has been made throughout the translation of Paré's text to preserve the informal flavor of his French, as well as the charm of his style. For this reason, the reader will note erratic spellings, irregular use of capitals, the presence of contractions, quaint vocabulary, and the like. Some explanations are offered in brackets to expedite comprehension; other longer explanations will be found in the notes. In addition, Paré's punctuation has been considerably modified in this edition to facilitate sentence flow.

Most of the scientific discussions arising here, both in the notes and in the Suggested Identifications, have been made by Philip D. Pallister, M.D., director of the Unit of Genetics and Birth Defects, Shodair Hospital, Helena, Montana, and William B. Jackson, D.Sc., University Professor in the Department of Biology, Bowling Green State University. Their contributions are identified by the initials following the entries in the notes and in the Suggested Identifications.

ON MONSTERS AND MARVELS

PREFACE

onsters are things that appear outside the course of Nature (and are usually signs of some forthcoming misfortune), such as a child who is born with one arm, another who will have two heads, and additional members over and above the ordinary.

Marvels are things which happen that are completely against Nature as when a woman will give birth to a serpent, or to a dog, or some other thing that is totally against Nature, as we shall show hereafter through several examples of said monsters and marvels, which examples I have gathered along with the illustrations from several authors, such as the *Histoires prodigieuses* of Pierre Boistuau, and from Claude Tiesserand, from Saint Paul, Saint Augustine, Esdras the Prophet, and from certain ancient philosophers, to wit from Hippocrates, Galen, Empedocles, Aristotle, Pliny, Lycosthenes, and others who will be quoted as deemed appropriate. 1

Maimed persons include the blind, the one-eyed, the humpbacked, those who limp or [those] having six digits on the hand or on the feet, or else having less than five, or [having them] fused together; or [having] arms too short, or the nose too sunken, as do the very flat-nosed; or those who have thick, inverted lips or a closure of the genitals in girls, because of the hymen; or because of a more than natural amount of flesh, or because they are hermaphrodites; or those having spots or warts or wens, or any other thing that is against Nature.

1

ON THE CAUSES OF MONSTERS

here are several things that cause monsters.

The first is the glory of God.

The second, his wrath.

The third, too great a quantity of seed.

The fourth, too little a quantity.

The fifth, the imagination.
The sixth, the narrowness or smallness of the womb.
The seventh, the indecent posture of the mother, as when, being pregnant, she has sat too long with her legs crossed, or pressed against her womb.
The eighth, through a fall, or blows struck against the womb of the mother, being with child.
The ninth, through hereditary or accidental illnesses.
The tenth, through rotten or corrupt seed.
The eleventh, through mixture or mingling of seed.
2 The twelfth, through the artifice of wicked spital beggars.
The thirteenth, through Demons and Devils.

(There are other causes that I leave aside for the present, because among all human reasons, one cannot give any that are sufficient or probable, such as why persons are made with only one eye in the middle of the forehead or the navel, or a horn on the head, or the liver upside down. Others are born having griffin's feet, like birds, and certain monsters which are engendered in the sea; in short countless others which it would take too long to describe. [1573]).

2

AN EXAMPLE OF THE GLORY OF GOD

aint John (Chapter 9) writes about a man who was born blind, who having recovered his sight, through the grace of Jesus Christ, the latter was interrogated by His disciples about whether his own sin or that of his parents was the cause of his having been brought forth blind from the very day of his birth. And Jesus Christ answered them [saying] that neither he nor his father nor his mother had sinned, but that it was in order that the works of God might be magnified in him.

3

AN EXAMPLE OF THE WRATH OF GOD

here are other causes which astonish us doubly because they do not proceed from the above-mentioned causes but from a fusing together of strange species, which render the creature not only monstrous but also to be marvelled at, that is to say, which is completely abhorrent and against Nature; as, why are some born with the form of a dog and the head of a fowl, another having four horns on the head, while another has the four feet of an ox and lacerated thighs, another having the head of a parrot and two plumes on the head and four talons, [and] others with other forms and configurations that you will be able to observe through several and diverse illustrations hereafter depicted according to their conformation.

It is certain that most often these monstrous and marvelous creatures proceed from the judgment of God, who permits fathers and mothers to produce such abominations from the disorder that they make in copulation, like brutish beasts, in which their appetite guides them, without respecting the time, or other laws ordained by God and Nature: as it is written in Esdras the Prophet (Ch. 5, Book 4), that women sullied by menstrual blood will conceive monsters.

Similarly, Moses forbids such coupling in Leviticus (Chapter 16). Also, the ancients observed through long experiences that the woman who will have conceived during her period will engender 3
those inclined to leprosy, scurvy, gout, scrofula, and more, or subject to a thousand different diseases: the more because a child conceived during the menstrual flow takes its nourishment and growth—being in its mother's womb—from blood that is contaminated, dirty, and corrupt, which having established its infection in the course of time, manifests itself and causes its malignancy to appear; some will have scurvy, others gout, others leprosy, others will have smallpox or measles, and endless other diseases. The conclusion is that it is a filthy and brutish thing to have dealings with a woman while she is purging herself.

1. Figure of a colt with a man's face

The aforementioned ancients estimated that such marvels often come from the pure will of God, to warn us of the misfortunes with which we are threatened, of some great disorder, and also that the ordinary course of Nature seemed to be twisted in [producing] such unfortunate offspring. Italy gave sufficient proof of this, for the travails she endured during the war that took place between the Florentines and the Pisans, after having seen in Verona, in the year 1254, a mare who foaled a colt which had the well-formed
4 head of a man, while the rest of him was a horse.

Another proof: From the time when Pope Julius II kindled so many misfortunes in Italy and when he waged war against King Louis XII (1512), which was followed by a bloody battle fought near Ravenna; just a little while afterwards, a monster was seen to be born having a horn on its head, two wings, and a single foot similar to that of a bird of prey, at the knee joint an eye, and participating in the *natures* [sexual organs] of both male
5 and female.

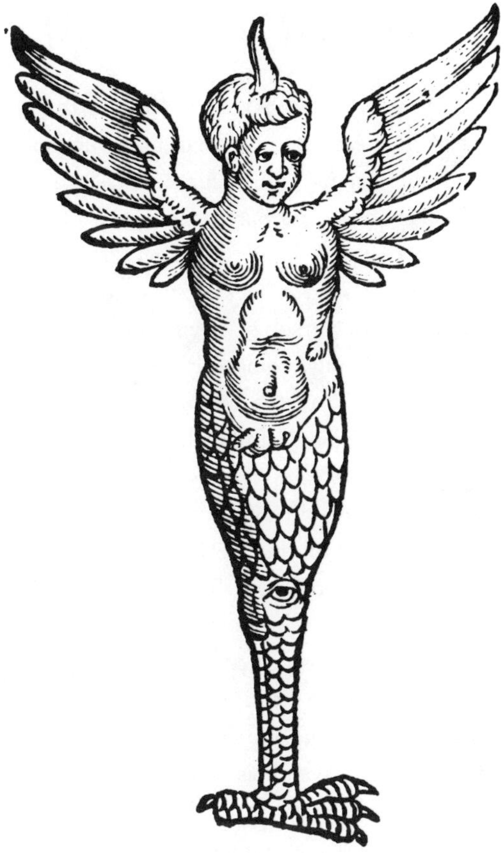

2. *Figure of a winged monster*

4

AN EXAMPLE OF TOO GREAT A QUANTITY OF SEED

On the generating of monsters, Hippocrates says that if there is too great an abundance of matter, multiple births will occur, or else a monstrous child having superfluous and useless parts, such as two heads, four arms, four legs, six digits on the hands and feet, or other things. And on the contrary, if the seed is lacking in quantity, some limb will be lacking, [such] as [a person's] having only one hand, no arms or feet or head, or [having] some other part missing.

Saint Augustine (Ch. 8 of the *City of God*) says that in his time
6 there was born in the Orient a child who had one belly; from there
upwards, all parts double, and downwards, all lower parts single,
for he had two heads and four eyes, two chests and four hands and
the rest like any other man, and he lived a rather long time.

Caelius Rhodiginus wrote in the book of his *Antiques leçons* (Ch.
7 3, Book 24) of having seen in Italy two monsters, one male and
the other female, their bodies perfect and well-proportioned, leaving
aside the duplication of the head; the male died a few days after
his birth, and the female, whose picture you see here [Fig. 3], lived
8 twenty-five years longer, which is not natural for monsters, who
ordinarily live scarcely any length of time at all because they grow
displeased and melancholy at seeing themselves so repugnant to
everyone, so that their life is brief.

Now one must note here that Lycosthenes (the great philosopher [1573]) writes a miraculous thing about this female monster, for leaving aside the duplication of the head, Nature had omitted nothing: these two heads (he says) had the same desire to drink, eat, sleep; and they had identical speech, as also their emotions were the same. This girl went begging from door to door for her livelihood, and people gladly gave to her on account of the novelty of such a strange and such a new spectacle. Nevertheless she was at last driven out of the Duchy of Bavaria because (they said) she could spoil the fruit of the pregnant women by the apprehension and ideas which might remain in their imaginative faculty, over the form of this so monstrous a creature. (It is not good that monsters should live among us. [1579])

In the year of grace 1475, there were similarly engendered in Italy, in the city of Verona, two girls joined together at the kidneys, from the shoulders clear to the buttocks[es]; and because their parents were poor, they were carted around to several cities in Italy, 9
in order to collect money from the people, who were burning to see this new spectacle of Nature. [Fig. 4] 10

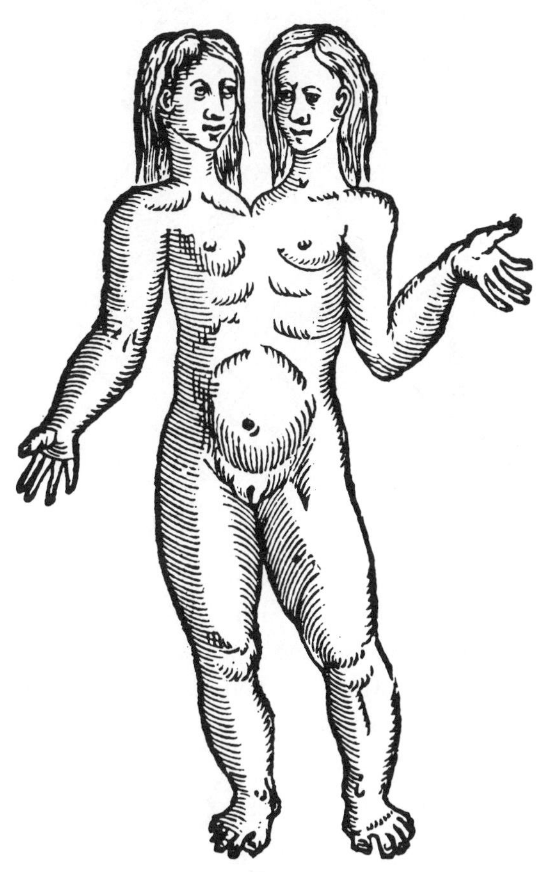

3. Figure of a girl having two heads

4. Figure of two twin girls joined and united by the posterior parts

In the year 1530, one could see in this very city of Paris a man from whose belly issued another man, well-formed in all his members, leaving aside the head, and this man was forty years old, or thereabouts, and he carried this body in his arms thus, so marvelously that people assembled in droves to see him, the figure of which is shown to you drawn from life.

In Piedmont, in the city of Quiers about five leagues from Turin a good woman gave birth to a monster on the seventeenth day of January at eight o'clock in the evening, in this very year of 1578, the face being well-proportioned in all its parts. It was found to be monstrous on the rest of the head, in that five horns approxi-

5. Figure of a man from whose belly another man issued

mating those of a ram came out of it, the horns being arranged one against the other on the top of the forehead and at the rear
a long piece of flesh hanging along the back, like a maiden's hood. 14
It had around its neck a flap of double-layered flesh like a shirt collar all of one piece, the extremities of the fingers resembling the talons of some bird of prey, its knees like hams. The foot and the right leg were an intense red color. The remainder of the body was a smoky gray color. It is said that at the birth of this monster it uttered a great bellow which so astonished the midwife and the whole crowd [gathered there] that the fright they had caused them to leave the building. Which news having reached his highness the

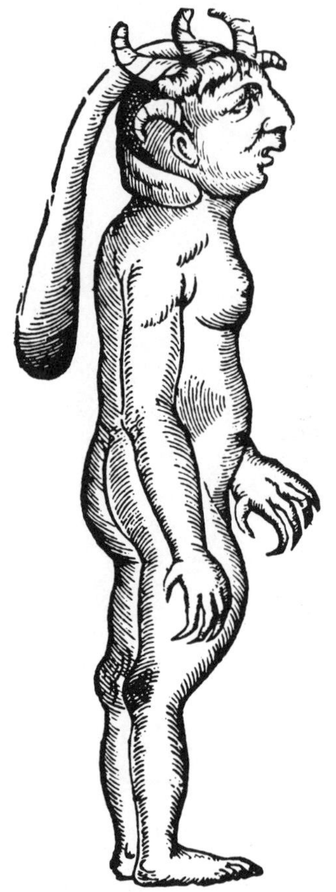

6. Figure of the horned or hooded monster

Prince of Piedmont, he sent for it, so much did he want to see it, in whose presence several persons pronounced various opinions; which monster is shown to you here in this true-to-life illustration.

The present monster that you see depicted here was found inside an egg, having the face and the visage of a man, his hair made up of little snakes—completely alive—and his beard in the mode and shape of three snakes which issued from his chin; and it was found on the fifteenth day of the month of March, last year, 1569, in the home of a counsellor-at-law named Baucheron, in Autun, in Bur-

7. The shape of a monster found in an egg

gundy, by a chambermaid who was cracking eggs in order to cook them in butter, among which was this one, which, being broken by her, she saw the aforesaid monster come out, having a human face, [but] the hair and beard made of live snakes, by which she was frightened and full of wonder. And the white of said egg was given to a cat, who instantaneously died from it. And M. le baron de Senecy, chevalier de l'ordre, when he was informed of this, had the aforementioned monster sent on his behalf to King Charles, who was then in Metz.

In the year 1546, in Paris, a woman who was six months pregnant gave birth to a child having two heads, two arms, and four legs,

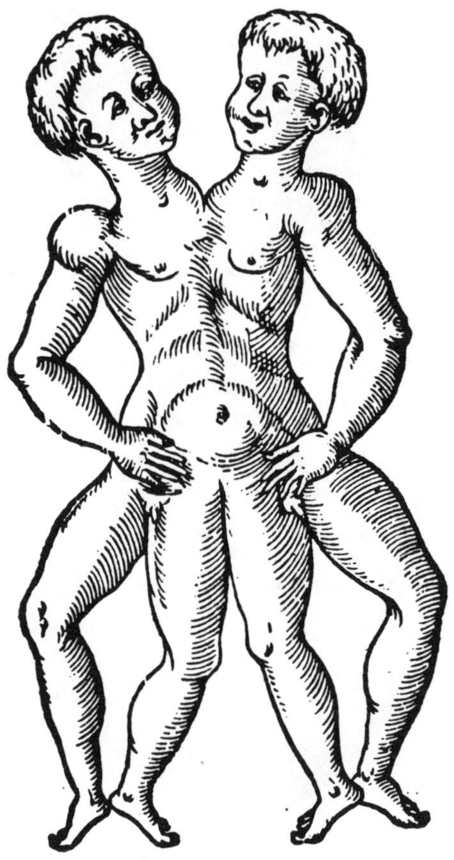

8. Figure of a child having two heads, two arms, and four legs

which I opened; and I found inside it only one heart (which monster is in my house and I keep it as [an example of] a monstrous thing), as a result of which one can say that it is only one child.

Aristotle (in his *Probl.*, and in Chap. 4, Book 4 of *Gener. animal.*) says that a monster having two bodies joined together, if it is found to have two hearts, one can truly say that there are two men or two women; otherwise, if it is found to have only one heart with two bodies, there is only one. The cause of this monster can have been a fault of matter, as to its quantity, or a defect in the womb,

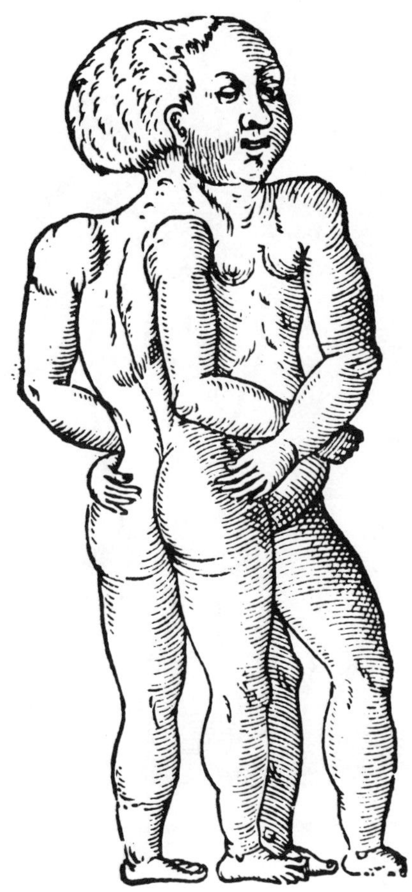

9. Figure of two twins having only one head

which was too small, because nature, wanting to create two children, finding it too small, finds itself lacking, so that the seed, being constrained and crowded, then comes to be coagulated in a globe, from which two children, thus joined and united, will be formed.

In the year 1569 a woman of Tours gave birth to two twin children having only one head, which were embracing each other; and they were given to me dry and dissected by master René Ciret, 16
master Barber and Surgeon, whose renown is widespread enough

10. Figure of two twin girls joined together by their foreheads

throughout the country of Touraine, without my having to give him any other praise. (These two last monsters are in the possession of the author.)

17 Sebastian Munster writes of having seen two girls, in the year 1495, in the month of September, near Worms, in the town named Bristant, who had bodies which were whole and well formed, but their foreheads were joined together, without anyone's being able to separate them by any human skill, and their noses almost touched each other; and they lived to the age of ten and then one died, who was lifted and separated from the other; and the one who remained alive died soon after, when her dead sister was separated

from her, because of the wound she had received from the separation; the illustration of which is shown to you here. 18

In the year 1570, on the twentieth day of July, in Paris, on the rue des Gravelliers, at the sign of the Bell, these two children shown below, were born. They were noted by the surgeons to be male and female and were baptized at St. Nicolas des Champs and named Louis and Louise. Their father's name was Pierre Germain, called Petit-Dieu, a mason's aid by trade, and their mother's name was Matthée Pernelle.

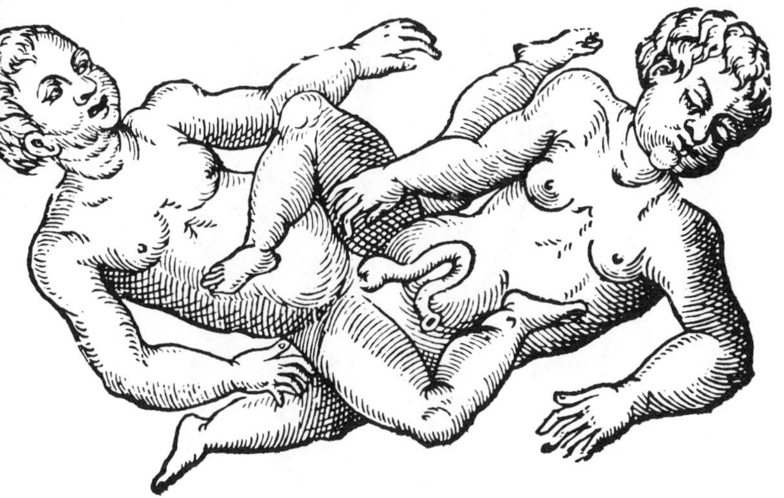

11. Figure of two monstrous children born not long ago in Paris

On Monday, the tenth day of July, 1572, in the city of Pont de Cé, near Angers, there were born two female children who lived for a half an hour and received baptism; and they were well-formed, except that one left hand had only four fingers, and they were joined together by their anterior parts, to wit, from the chin to the umbilicus, and had only one navel, and a single heart, the liver being divided into four lobes.

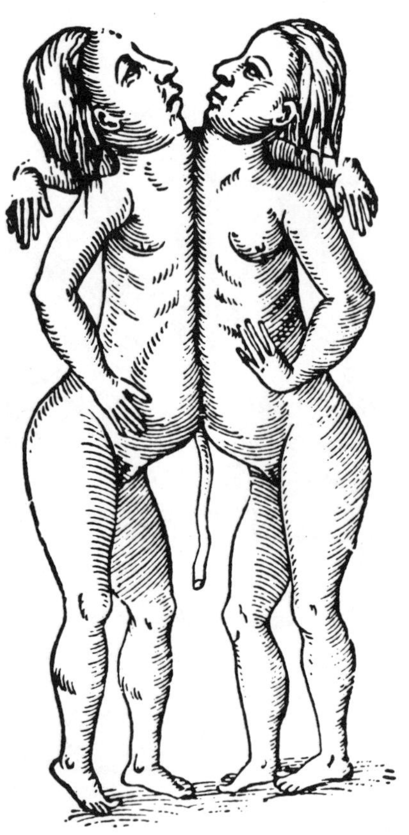

12. Figure of two girls joined together by their anterior parts

Caelius Rhodiginus, in the third chapter, the twenty-fourth book of his *Antiques leçons,* writes that a monster was produced in Ferrara, in Italy, in the year of grace 1540, the nineteenth day of March, who, when he was delivered, was as big and well-formed as if he were four months old, having both feminine and masculine sexual organs, and two heads, the one of a male and the other of a female. 19

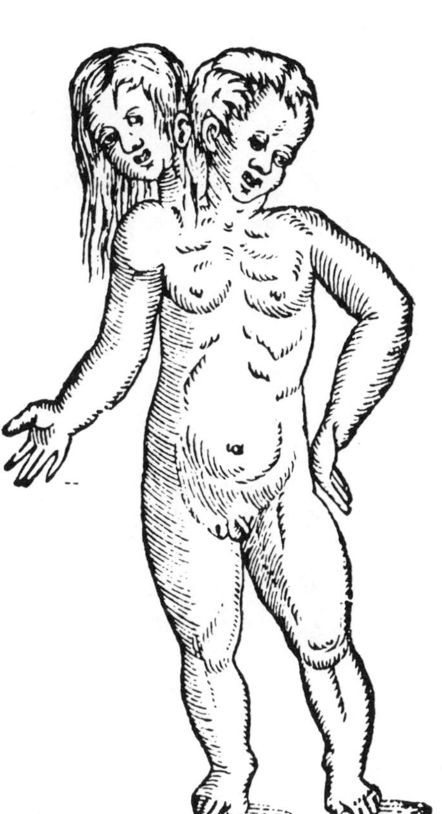

13. Figure of a monster having two heads, one male and the other female

Jovianus Pontanus writes that in the year 1529, the ninth of January, there was seen in Germany a male child having four arms and four legs, whose picture you see here.

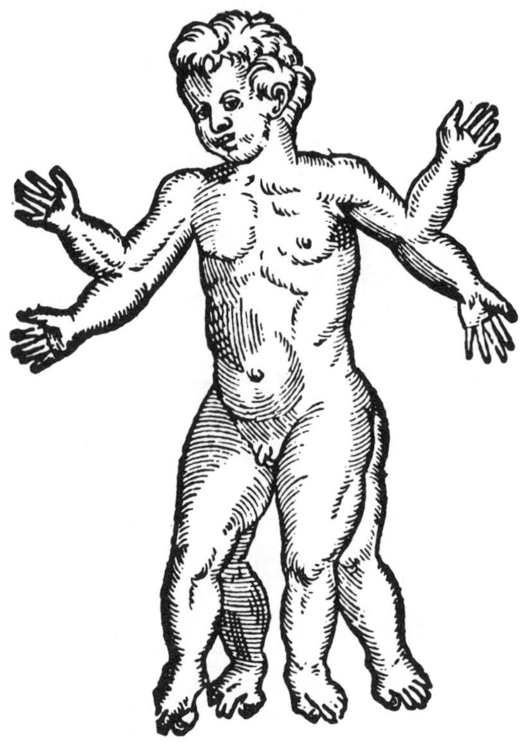

14. Figure of a monster with four legs and four arms

In the same year that the great king Francis [I] made peace with the Swiss, a monster was born in Germany having a head in the middle of its stomach; he lived to be an adult; and the head took nourishment just like the other. 21

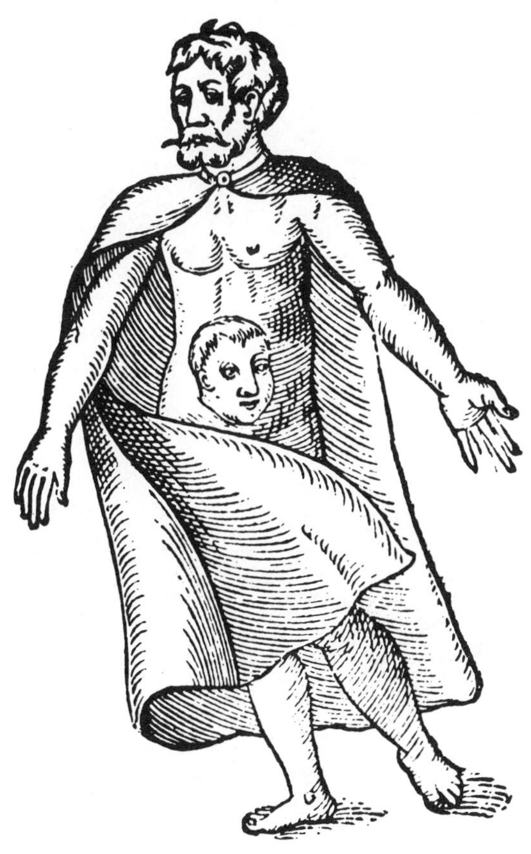

15. Figure of a man having a head in the middle of his belly

On the last day of February, 1572, in the parish of Viabon, on the road from Paris to Chartres, at the place of the Petites Bordes, a woman named Cypriane Girande, the wife of Jacques Marchant, a tiller, delivered this monster, who lived until the following Sunday. [Fig. 16] 22

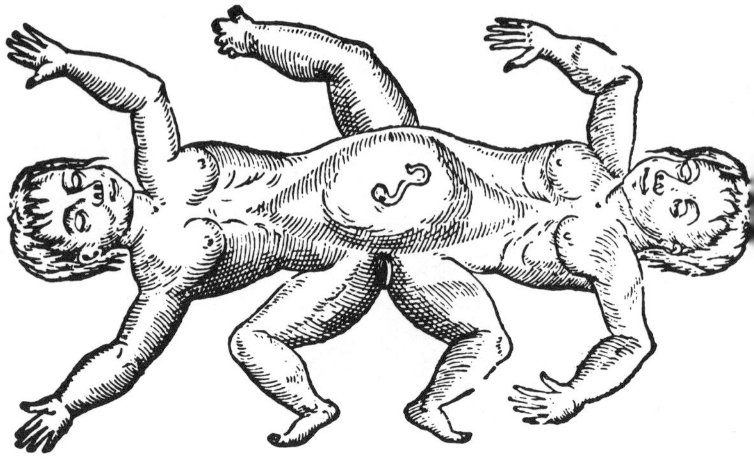

16. Figure of two very monstrous infants, in whom only one set of female sexual organs is manifested

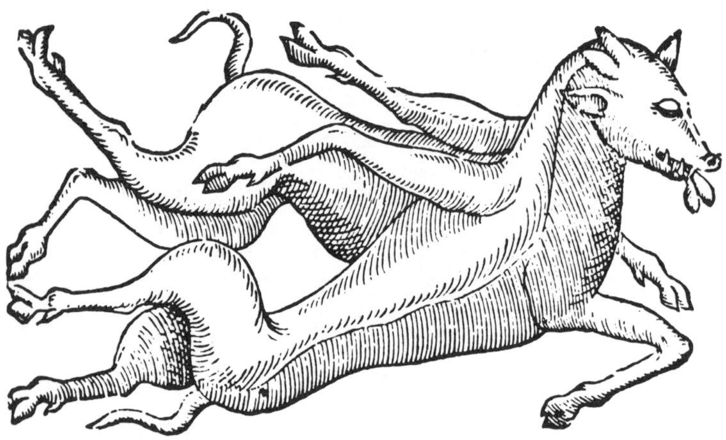

17. Figure of a monstrous pig, born in Metz in Lorraine

In the year 1572, the day after Easter, in Metz, in Lorraine, in the inn of the Holy Spirit, a sow farrowed a pig having eight legs, four ears, the head of a real dog, the hind quarters of the bodies separated up to the stomach and from there on joined into one, having two tongues located on the wrong side of the snout; and it had four large teeth, to wit above as well as below, on either side. Their sexual organs were not well-defined, so that one could not tell if they were male or female. Each one had only one vent beneath the tail. The form of this monster is depicted for you by the preceding picture which was sent to me not long ago by monsieur Bourgeois, Doctor of Medicine, a very learned man, and very experienced in that field, living in the aforementioned city of Metz.

At this spot it seems to me not to be off the subject to write about women who carry several children during one pregnancy.

5

ON WOMEN WHO CARRY SEVERAL CHILDREN DURING ONE PREGNANCY

Women commonly deliver one child; however, one can see—as there are a great many women in the world—that they *may* deliver two, which are called twins or "bessons"; there some women who deliver three, four, five, six, and more.

Empedocles says that when there is a great quantity of seed, a plurality of children is produced. Others, such as the Stoics, say that they are conceived because in the womb there are several cells, separations, and cavities, and when the seed is spread into them several children are created. However, that is false, for in the womb of women only one cavity alone is to be found, but in animals, such as bitches, swine, and others, there are several cells, which is the cause of their carrying several babies.

Aristotle has written that women could not deliver more than five children at one pregnancy; yet it happened to the woman-servant of Augustus Caesar that at one pregnancy she delivered five children, which (along with the mother) lived only a short while.

In the year 1554 in Bern, Switzerland, the wife of John Gislinger, Doctor, likewise delivered five children from one pregnancy, three
23 males and two females.

24 Albucrasis says he is certain concerning a lady who delivered seven of them and concerning another, who, having injured herself aborted fifteen well-formed ones. Pliny (Chap. 11, Book 7) mentions one who aborted twelve of them. The same author says that at Peloponnesus a woman was seen who delivered four times, and at each pregnancy she had five children, most of whom lived.

Dalechamps, in his *French Surgery* (Ch. LXXIV, folio 118), says that a gentleman named Bonaventure Savelli, of Siena, assured him that a slave of his, whom he retained, had seven children in one pregnancy, four of whom were baptized. And in our own time between Sarte and Maine, in the parish of Seaux, near Chambellay there is a gentleman's house called Maldemeure, whose wife had
25 two children the first year she was married, three the second year four the third year, five the fourth year, six the fifth year, from which she died: there is one of the aforementioned six children [still] living, who today is Sire [or, Lord] of the aforementioned Maldemeure.

In Beaufort in the valley, in the country of Anjou, a young woman, the daughter of the late Macé Chaumière, delivered a child and eight or ten days later another one, that they had to extract from her womb, from which she died.

26 Martinus Cromerus, in Book 9 of his *History of Poland,* writes
27 that, in the province of Krakow, Marguerite, a very virtuous lady and of a great and ancient house, the wife of a count called Virboslaus, delivered on the twentieth day of January, 1269, a litter of thirty-six children born alive.

28 Franciscus Picus Mirandula writes that a woman in Italy, named Dorothea, gave birth to twenty children in two confinements, namely nine children at one time and eleven at another, who, carrying such a great load, was so heavy that she held up her stomach, which

18. The picture of Dorothy, pregnant with many children

hung down as far as her knees, with a great hoop which hung over her neck and shoulder, as you see in this picture. 29

Now, as for the reason for multiple births, some persons, completely ignorant of anatomy, have tried to argue that in the womb of the woman there were several cells and sinuses, namely seven of them, three on the right side for males, three on the left for females, and the seventh one right in the middle for hermaphrodites; and this untruth has been authorized even to the point that some persons afterward have affirmed each one of these cavities to be divided yet again into ten others, and from this they have concluded that multiple births in one litter are due to the fact that diverse portions of the seed were scattered and received in several cells. But such a thing is not supported by reason or by any authority; rather it is contrary to sense and observation, although Hippocrates seems to have been of this opinion in the book *De natura pueri;* but Aristotle, book four, chapter four, *De generatione animal.,* thinks that twins or multiple births in one litter are formed in the same way as a sixth finger on the hand, namely, through the superfluity of matter, which, being in great abundance, if it reaches the point of dividing in two, twins are formed.

It has seemed to me to be a good idea that I should describe hermaphrodites at this point because they also come from a superabundance of matter.

6

ON HERMAPHRODITES OR ANDROGYNES, THAT IS TO SAY, WHICH HAVE TWO SETS OF SEX ORGANS IN ONE BODY

ermaphrodites or androgynes are children who are born with double genitalia, one masculine and the other feminine, and as a result are called in our French language "hommes et femmes" [men-and-women]. (Androgyne in Greek means man and woman, and woman and man.)

Now as for the cause, it is that the woman furnished as much seed as the man proportionately, and for this [reason] the formative virtue [property], which always tries to make its likeness—to wit, a male from the masculine matter and a female from the feminine—operates so that sometimes two sexes, called hermaphrodites, are found in the same body. Of these, there are four different kinds, to wit, the male hermaphrodite, which is the one that has the sex organs of the perfect man and can impregnate and has, at the

perineum (which is the spot beneath the scrotum and the seat), a 30
hole in the form of a vulva, which nonetheless does not penetrate to the inside of the body; and urine and seed do not issue from this hole. The hermaphroditic woman, in addition to her vulva which is well-formed and through which she ejects seed and her monthlies, has a male member, situated above the said vulva, near the groin, without foreskin, but with a thin skin which cannot be turned over or around, and without any erection, and from this, urine and seed do not issue; and no vestige of scrotum or testicles is to be found there. The hermaphrodites who are neither one nor the other [of the aforementioned] are those who are totally excluded from and void of reproduction, and their sexual organs are wholly imperfect and are situated alongside one another and sometimes one on top and the other beneath, and they can use them only to eject urine. Male and female hermaphrodites are those who have both sets of sexual organs well-formed, and they can help and be used in reproduction; and both the ancient and modern laws have obliged and still oblige these latter to choose which sex organs they wish to use, and they are forbidden on pain of death to use any but those they will have chosen, on account of the misfortunes that could result from such. For some of them have abused their sit- 31
uation, with the result that, through mutual and reciprocal use, they take their pleasure first with one set of sex organs and then with the other: first with those of a man, then with those of a woman, because they have the *natures* of man and of woman suitable to such an act; verily, as Aristotle describes (in his *Probl.*, sec. on *Hermaphrodites*, pro. 3 and 4), their right teat is just like that of a man and their left one like that of a woman.

The most expert and well-informed physicians and surgeons can recognize whether hermaphrodites are more apt at performing with and using one set of organs than another, or both, or none at all. And such a thing will be recognized by the genitalia, to wit, whether the female sex organ is of proper dimensions to receive the male rod [penis] and whether the menstrues flow through it; similarly by the face and by the hair, whether it is fine or coarse; whether the speech is virile or shrill; whether the teats are like those of men or of women; similarly whether the whole disposition of the body is robust or effeminate; whether they are bold or fearful, and other actions like those of males or of females. And as for the genitalia which belong to a man, one must examine to see whether there is a good deal of body hair on the groin and around the seat, for commonly and almost always women have none on the seat. Similarly, one must examine carefully to see whether the male rod

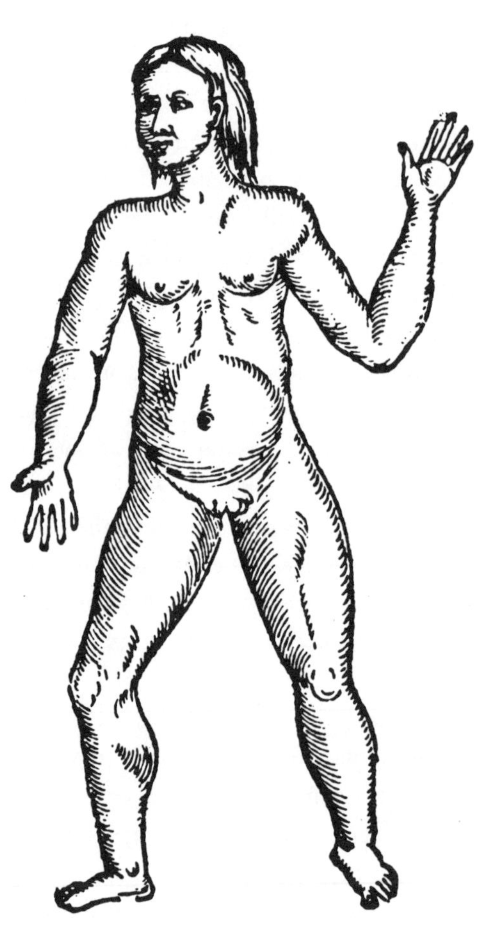

19. Picture of a hermaphrodite man-and-woman

s well-proportioned in thickness and length, and whether it can [become] erect, and whether seed issues from it, [all of] which will be done through the confession of the hermaphrodite, when he will have kept company with a woman; and by this examination one will truly be able to discern and know the male or female hermaphrodite, either that they will be one and the other, or that they will be neither one nor the other. And if the sexual organs of the hermaphrodite are more like those of a man than of a woman, he

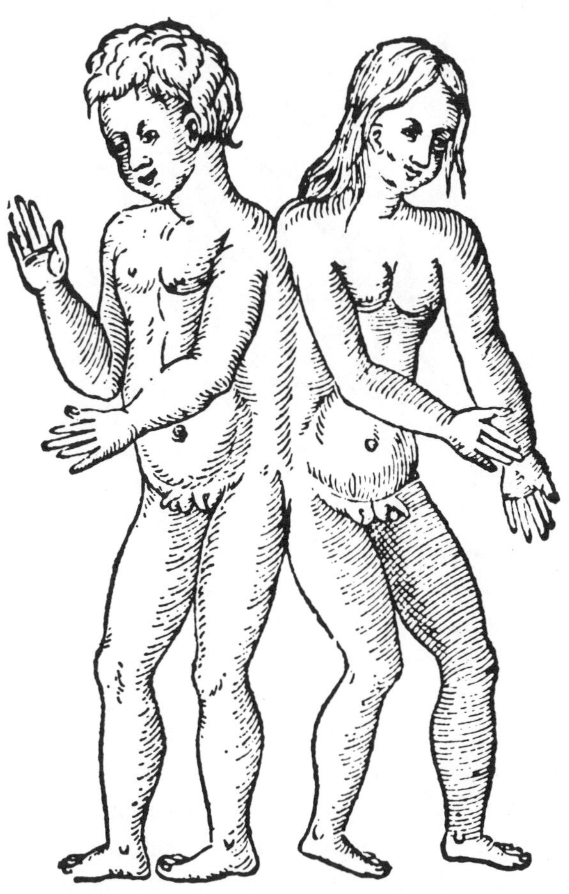

20. Figure of two hermaphroditic twin children, being joined back to back the one to the other

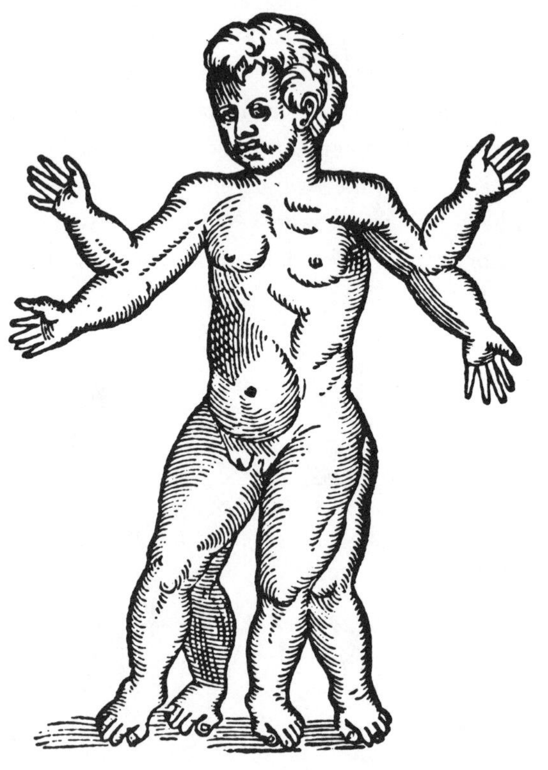

21. Figure of a monster having four arms and four feet and two female "natures"

is to be called a man; and the same will be true for a woman. And
if the hermaphrodite is as much like one as another, he will be
32 called a hermaphrodite, or a "man-and-woman." [Fig. 19]
33 The year 1486 saw the birth in the Palatinate, rather near
Heidelberg, in a burg named Rorbarchie, of two twin children
holding each other up and joined together back to back, who were
34 hermaphrodites, as one can see by this figure. [Fig. 20]

On the day that the Venetians and Genevans were reconciled there was born in Italy (as Boistuau recounts) a monster which had four arms and four legs; and it had only one head, with the [correct] proportion conserved in all the rest of the body, and it was baptized and lived for some time afterward.

Jacques Rueff, a surgeon of Zurich, writes that he had seen a similar one, which had two female *natures,* as you can see by this picture. 35

7
MEMORABLE STORIES ABOUT WOMEN WHO HAVE DEGENERATED INTO MEN

Amatus Lusitanus tells that there was, in a burg 36
named Esgueira, a girl called Marie Pacheca, who, arriving at the time of life when girls begin their monthlies, instead of the above-mentioned monthlies, a male member came out of her, which was formerly hidden within, and hence she changed from female to male; for which reason she was clothed in men's clothes and her name was changed from Marie to Manuel. This male person traded for a long time in India, where, having acquired great fame and great wealth, upon returning [to Portugal] got married; nevertheless, this author [Lusitanus] doesn't know whether he had children; it is true (says he) that he remained without a beard.

Antoine Loqueneux, tax [or rent] receiver for the King in Saint
Quentin, not long ago assured me he had seen a man in the Inn 37
of the Swan in Reims, sixty years of age, who, so it seems, people had taken to be a girl until he was fourteen; but while disporting himself and frolicking, having gone to bed with a chambermaid, his male genital parts came to be developed; the father and the mother, recognizing him to be such, had him by authority of the Church change his name from Jeanne to Jean, and male attire was given to him.

Also being in the retinue of the King at Vitry-le-François in
Champagne, I saw a certain person (a shepherd) named Germain 38
Garnier—some called him Germain Marie, because when he had been a girl he had been called Marie—a young man of average

size, stocky, and very well put together, wearing a red, rather thick beard, who, until he was fifteen years of age, had been held to b a girl, given the fact that no mark of masculinity was visible i him, and furthermore that along with the girls he even dressed lik a woman. Now having attained the aforestated age, as he was i the fields and was rather robustly chasing his swine, which wer going into a wheat field, [and] finding a ditch, he wanted to cros over it, and having leaped, at that very moment the genitalia an the male rod came to be developed in him, having ruptured th ligaments by which previously they had been held enclosed an locked in (which did not happen to him without pain), and, weeping he returned from the spot to his mother's house, saying that hi guts had fallen out of his belly; and his mother was very astonishe at this spectacle. And having brought together Physicians an Surgeons in order to get an opinion on this, they found that sh was a man, and no longer a girl; and presently, after having reporte to the Bishop—who was the now defunct Cardinal of Lenoncort— and by his authority, an assembly having been called, the shepher received a man's name: and instead of Marie (for so was h previously named), he was called Germain, and men's clothing wa given to him; and I believe that he and his mother are still living

Pliny (Book 7, Chapter 4) says similarly that a girl became a boy and was for this reason confined on a deserted and uninhabite island, by the decision and order of the Aruspices (or soothsayers) It seems to me that these prophets had not any cause to do this for the reasons given above; still they estimated that such a monstrous thing was a bad augury and presage for them, which wa the reason for driving monsters away and exiling them.

The reason why women can degenerate into men is because wome have as much hidden within the body as men have exposed outside leaving aside, only, that women don't have so much heat, nor th ability to push out what by the coldness of their temperament is hel as if bound to the interior. Wherefore if with time, the humidity o childhood which prevented the warmth from doing its full duty bein exhaled for the most part, the warmth is rendered more robust vehement, and active, then it is not an unbelievable thing if the latter chiefly aided by some violent movement, should be able to push out what was hidden within. Now since such a metamorphosis takes place in Nature for the alleged reasons and examples, we therefore never find in any true story that any man ever became a woman, because Nature tends always toward what is most perfect and not, on the contrary, to perform in such a way that what is perfect should become imperfect. 39

8

AN EXAMPLE OF LACK IN THE QUANTITY OF SEED

f the quantity of seed (as we said prior to this) is deficient, [then] similarly one or more members will also be lacking.

From this it will result that one child will have two heads and one arm, and another will have no arms at all; another will have neither arms nor legs, or other parts [will be] lacking, as we said above; another will have two heads and only one arm, and the rest of the body will be complete, as you see from this figure. [Fig. 22] 40

In the year 1573 I saw in Paris, at the door of Saint Andrew of the Arts, a child about nine years old, a native of Parpeville, a village three leagues from Guise; its father was named Pierre Renard and the mother who bore it Marquette. This monster had only two fingers on its right hand, and the arm was rather well-formed from the shoulder to the elbow, but from the elbow to the two fingers [it] was very deformed. It was without legs, and yet there issued from his right buttock the incomplete form of a foot, appearing to have four toes; from the left buttock there issued from the middle two toes, one of which resembled, almost, the male rod. Which is shown to you just as it was through this present picture. [Fig. 23]

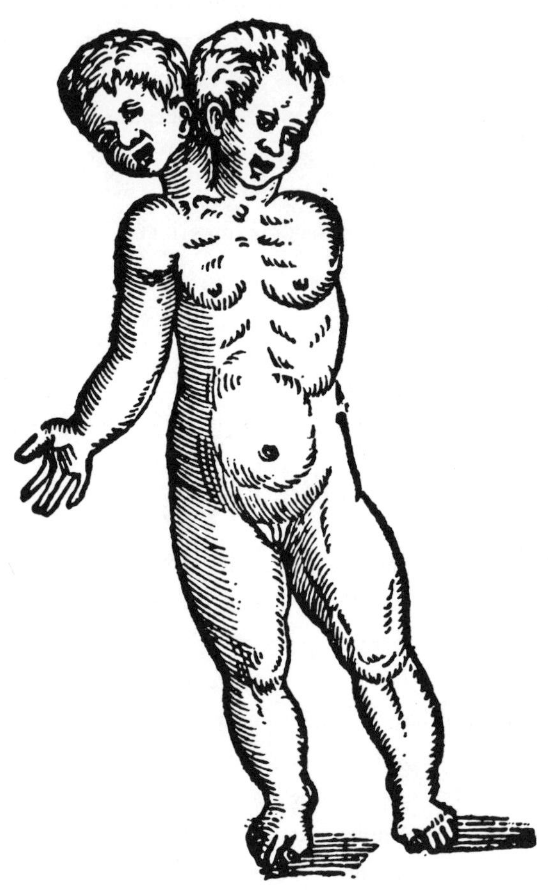

22. *Figure of a monster having two heads, two legs, and only one arm*

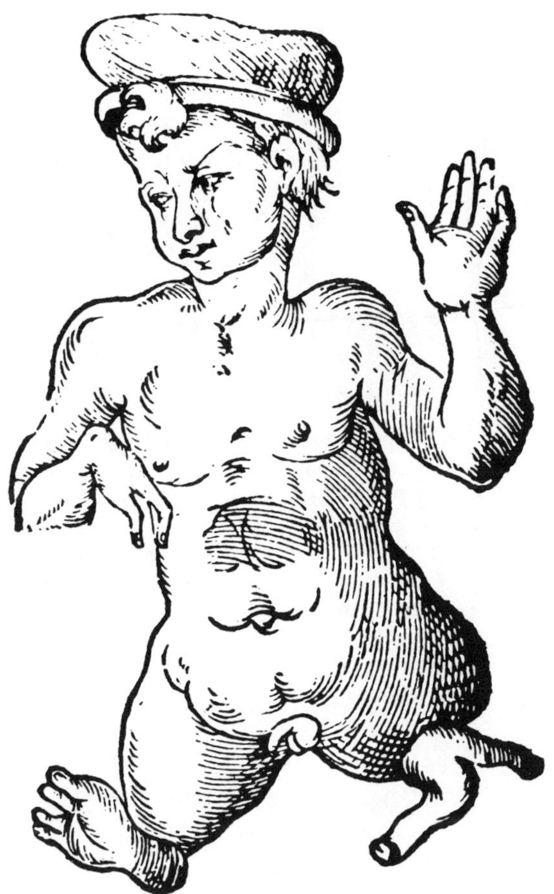

23. Figure of a monstrous child, coming from a lack of seed in proper quantity

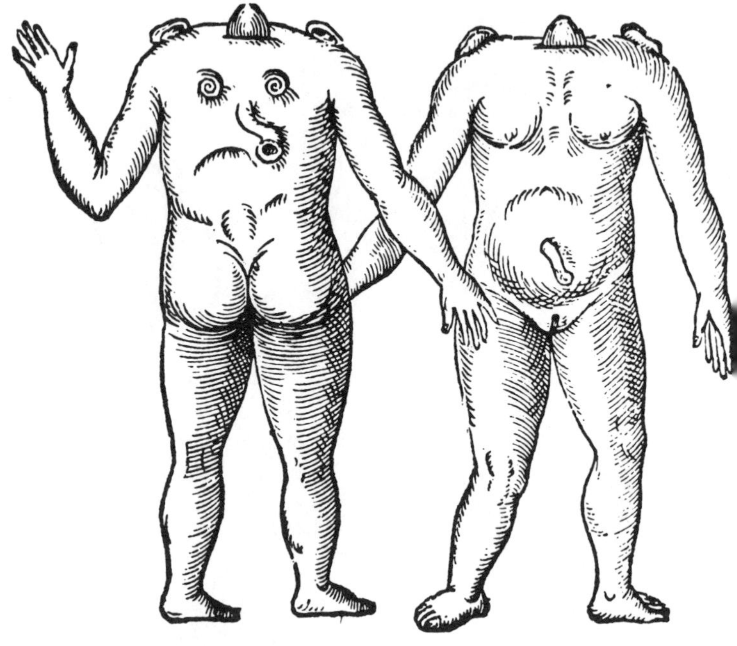

24. Figure of a female monster without a head, front and back views

In the year 1562, on the first day of November, there was born
41 in Ville-franche-de-Beyran in Gascony, this present monster, without a head, which was given to me by Monsieur Hautin, regent doctor of the faculty of medicine in Paris, of which monster you have here a picture, both of the anterior and of the posterior, and
42 he assured me that he had seen it.

Some while ago one could see in Paris a man without any arms, forty years old, or thereabouts, strong and robust, who performed almost all the actions that another might do with his hands; to wit, with his stump of a shoulder and his head he would strike a hatchet against a piece of wood as firmly as another man might have been able to do with his arms. Similarly, he could make a carter's whip snap, and he performed several other actions; and with his feet he ate, drank, and played cards and dice, which is shown to you in this picture. In the end, he was a robber, thief, and murderer and

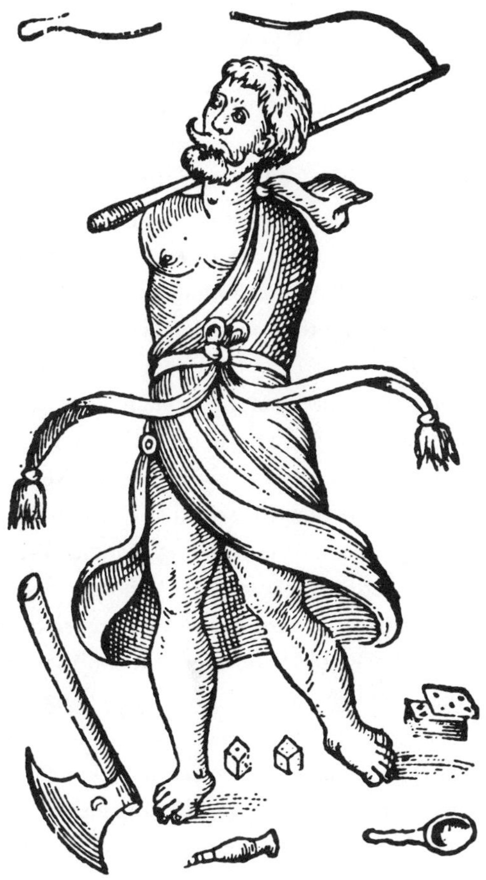

25. Figure of a man without arms

was executed in Guelders, that is [to say], hanged and then fastened to a wheel. 43

Similarly, of recent memory, an armless woman was seen in Paris who cut cloth and sewed and performed several other actions.

Hippocrates in Book 2 of *On Epidemics* writes that the wife of Antigenes gave birth to a child which was entirely of flesh, having no bones; nevertheless all its parts were well-formed. 44

9

AN EXAMPLE OF MONSTERS THAT ARE CREATED THROUGH THE IMAGINATION

he ancients, who sought out the secrets of Nature (i.e., Aristotle, Hippocrates, Empedocles), have taught of other causes for monstrous children and have referred them to the ardent and obstinate imagination [impression] that the mother might receive at the moment she conceived—through some object, or fantastic dream—of certain nocturnal visions that the man or woman have at the hour of conception. This is even verified by the authority of Moses (Chap. 30 [of Genesis]) when he shows how Jacob deceived his father-in-law Laban and enriched himself with his livestock by having rods barked and putting them in the watering trough, so that when the goats and ewes looked at these rods of various colors, they might form their young spotted in various colors: because the imagination has so much power over seed and reproduction that the stripe and character of them remain [imprinted] on the thing bred.

Whether true or not, Heliodorus (book 10, of his *History of Ethiopia*) writes that Persina, the Queen of Ethiopia, conceived by King Hidustes—both of them being Ethiopians—a daughter who was white and this [occurred] because of the appearance of the beautiful Andromeda that she summoned up in her imagination, for she had a painting of her before her eyes during the embraces
45 from which she became pregnant.

46 Damascene, a serious author, attests to having seen a girl as furry as a bear, whom the mother had bred thus deformed and hideous, for having looked too intensely at the image of Saint John [the Baptist] dressed in skins, along with his [own] body hair and beard, which picture was attached to the foot of her bed while she was conceiving.

47 For a similar reason Hippocrates saved a princess accused of
48 adultery, because she had given birth to a child as black as a Moor, her husband and she both having white skin; which woman was

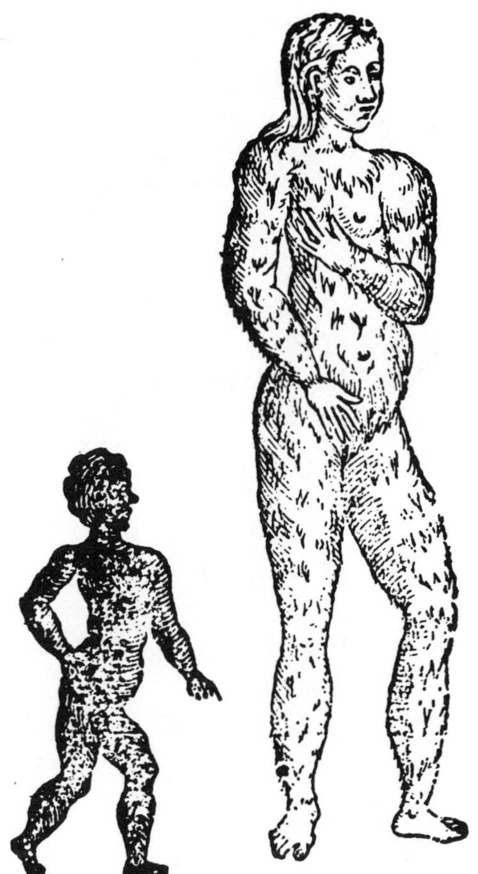

26. *Two figures, one of a furry girl, and the other of a child that was black because of the imagination of their parents*

absolved upon Hippocrates' persuasion that it was [caused by] the portrait of a Moor, similar to the child, which was customarily attached to her bed.

Moreover, one can observe that conies [rabbits] and peacocks who are closed up in white places, through the properties of the imagination, give birth to their white young.

As a result, it is necessary that women—at the hour of conception and when the child is not yet formed (which takes from

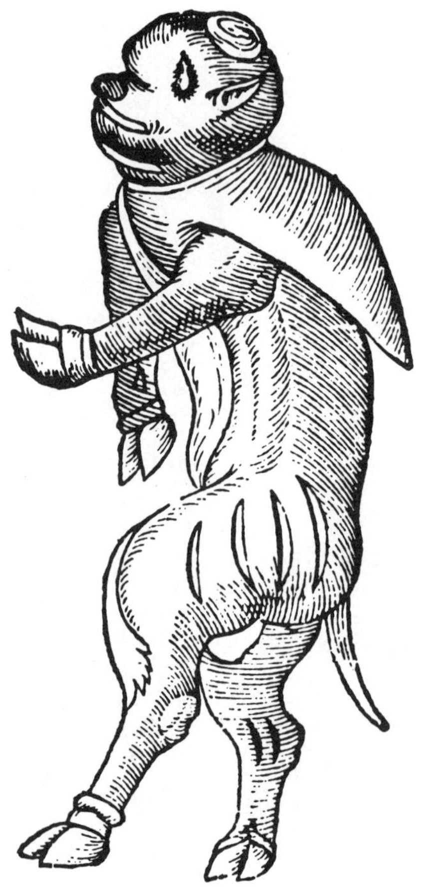

27. Figure of a very hideous monster having the hands and feet of an ox, and other very monstrous things

thirty to thirty-five days for males and forty or forty-two, as Hippocrates says, for females)—not be forced to look at or to imagine monstrous things; but once the formation of the child is complete, even though the woman should look at or imagine monstrous things with intensity, nevertheless the imagination will not then play any role, because no transformation occurs at all, since the child is completely formed.

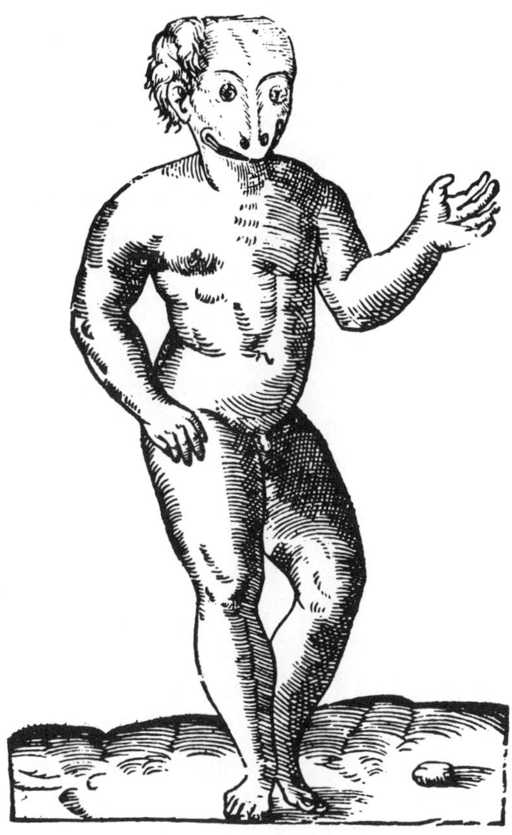

28. Prodigious figure of a child having the face of a frog

In Saxony in a village named Stecquer, a monster was born 49
having the four feet of an ox; its eyes, mouth and nose similar to
a calf, having on top of its head a red flesh, in a round shape;
[and] another behind, similar to a monk's hood, and having its
thighs mangled. 50

In the year 1517, in the parish of Bois-le-Roy, in the Forest of Bière, on the road to Fontainebleau, a child was born having the face of a frog, who was seen and visited by Master Jean Bellanger, a surgeon in the company of the King's Artillery, in the presence of gentlemen from the Court of Harmois: notably the honorable

gentleman Jacques Bribon, the king's procurer in said place; and Etienne Lardot, a bourgeois from Melun; and Jean de Vircy, king's notary at Melun; and others. The father's name is Esme Petit and the mother Magdaleine Sarboucat. The aforementioned Bellanger, a man of good wit, wanting to know the cause of the monster, inquired of the father what could have been the cause of it; the latter told him that he figured that, his wife having a fever, one of her neighbor ladies advised her, in order to cure her fever, to take a live frog in her hand and hold it until said frog should die. That night she went to bed with her husband, still having said frog in her hand; her husband and she embraced and she conceived; and by the power of her imagination, this monster had thus
51 been produced.

10

AN EXAMPLE OF THE NARROWNESS OR THE SMALLNESS OF THE WOMB

onsters are also created through the extreme narrowness of the womb, just as one sees that when a pear attached to the tree and put into a narrow vessel
52 before it is completely grown cannot attain full growth; which is also known to ladies who foster puppies in small baskets or other narrow containers in order to keep them from growing. Similarly, a plant being born in the earth, finding a rock or other solid thing in the spot where it is growing, causes the plant to be twisted, and thick in one part and thin in another: likewise, infants come out of the womb of their mothers monstrous and deformed. For he says (Hippocrates, in his book *On Reproduction*) that it is inevitable that a body which moves in a small place should become mutilated and lacking.

53 Empedocles and Diphilus have similarly attributed this to superabundance, or [to] lack and corruption of the seed, or to the

indisposition of the womb; which may be [seen to be] true, through its resemblance to fusible things, in which—if the substance that one wishes to melt is not well cooked, purified, and prepared or if the mold should be rough or in some other way badly disposed—the medal or effigy that issues from it it is defective, hideous, and deformed.

11

AN EXAMPLE OF MONSTERS THAT ARE FORMED, THE MOTHER HAVING REMAINED SEATED TOO LONG, HAVING HAD HER LEGS CROSSED, OR HAVING BOUND HER BELLY TOO TIGHT WHILE SHE WAS PREGNANT

ow sometimes it also happens by accident that the womb is by nature ample enough, but that even so, the woman being pregnant, on account of having almost always held her legs crossed during pregnancy—as seamstresses and women who work tapestries on their knees often do quite readily—or having bound or strapped her belly too tight, the children are born bent, hunchbacked and misshapen, some having
their hands and feet twisted as you see by this figure. [Fig. 29] 54

[Below is a] portrait of a marvel, a putrefied child which was found in the body of a dead woman in the city of Sens on the sixteenth day of May, 1582, she being sixty-eight years old and having carried it in her womb for the space of twenty-eight years. Said child was almost completely gathered up into a ball, but it is depicted here at its full length, in order to show better the entire conformation of its members, except for one hand, which was defective.

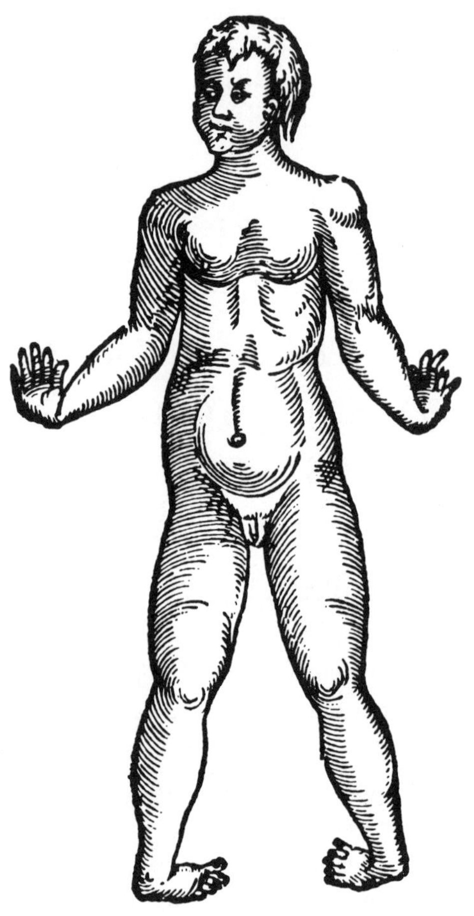

29. Figure of a child who has been pressed against the mother's belly, having his hands and feet twisted

This [kind of monstrosity] can be confirmed by Matthias Cornax, physician of Maximilian, king of the Romans, who narrates that he himself was present at the dissection of the womb of a woman who had carried her child in her womb for the space of four years. Also, Egidius Hertages, a physician in Brussels, mentions a woman

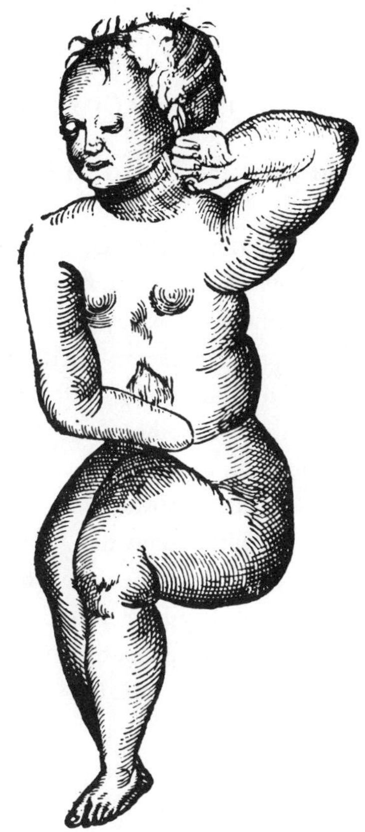

30. [Figure of dead fetus carried in mother's womb for twenty-eight years]

who carried in her flanks for thirteen whole years the skeleton of a dead child. Joannes Langius, in the epistle he writes to Achilles Bassarus—bears witness also of a woman who was from a burg called Eberbach [and] who spewed the bones of a child who had died in her womb ten years before.

12

AN EXAMPLE OF MONSTERS WHO ARE CREATED, THE MOTHER HAVING RECEIVED SOME BLOW OR FALL, BEING GREAT WITH CHILD

Moreover, when the mother receives some blow over the womb or when she falls down from a certain height, the infants can have their bones broken, thrown out of joint and twisted, or receive some other defect, such as being crippled, hunchback and misshapen; or for the reason that the child becomes sick in the womb of its mother, or because the nourishment which he was to receive to grow on flowed out of the womb. Similarly, some have attributed monsters to being procreated from the corruption of foul and filthy foods that women eat, or want to eat, or that they abhor looking upon just after they have conceived; or [they say] that someone may have tossed something between their teats, such as a cherry, plum, frog, mouse, or other thing that can render infants monstrous.

13

AN EXAMPLE OF MONSTERS THAT ARE CREATED BY HEREDITARY DISEASES

Also, because of hereditary indispositions or compositions of the fathers and mothers, children are made monstrous and deformed: for it is rather manifest that a hunchback gives birth to a hunchbacked child, indeed so hunchbacked that the two humps before and behind on some of them are so very elevated that the head is half-hidden between the
56 shoulders, just like the head of a tortoise in its shell. A woman who drags one side of her body grows children who limp the same way she does; others being crippled in both hips grow children who

are that way and who make their way by waddling like ducks; very
flat-nosed persons grow very flat-nosed children; others stammer,
while still others speak in a garbled way, and similarly their children
speak in a garbled way. (*Balbutier,* to stammer, that is to say, stutter,
not being able to pronounce a word. *Bredouiller,* to speak in a
garbled way, is to say a word two or three times without pronouncing
it well.) And when the fathers and mothers are small, children born 57
to them are usually dwarfs, without any other deformity, to wit,
when the body of the father and mother have no defect in their
conformation. Other persons grow very thin children, because the
father and mother are so; others are big-bellied or have large
buttocks[es], almost thicker than they are long, because they have
been bred by a father or a mother, or both, who are fat and tall,
big-bellied and having large buttocks[es]. Gouty people engender
gout in their children and *lapidaires* [create children] subject to 58
stones; also if the father and the mother are fools usually the
children are scarcely if ever intelligent (similarly epileptics give
birth to children who are subject to epilepsy). 59

Now all these kinds of people are far from unusual, which is
a thing that anyone can observe and [therefore one can] know by
his own eye the truth of what I am saying, wherefore I see no
reason to continue talking further about it. In addition, I have no
desire to write that lepers give birth to leprous children, for everybody knows this. There is an infinity of other dispositions in fathers
and mothers to which their children are subject; even manners and
speech, their mien and visage, countenance and gesture, even their
way of walking and spitting. Nonetheless, one must not make a
definite rule out of this, for we see fathers and mothers who have
any one of these indispositions and yet the children retain nothing
of it because the formative power [or property] has
corrected this defect. 60

14

AN EXAMPLE OF MONSTROUS THINGS WHICH HAVE OCCURRED IN ACCIDENTAL ILLNESSES

ust outside of Saint Jean d'Angely a soldier named Francisque, of Captain Muret's company, was wounded by a handgun shot to the belly, between the
61 umbilicus and the "Isles." The bullet was not extracted from him, because they couldn't find it, as a result of which he had great and extreme pain; nine days after receiving his wound, he emitted the bullet through the seat, and three weeks afterward he was cured; he was treated by master Simon Crinay, Surgeon of the French troops.

Jacques Pape, Lord of Saint-Aubam-aux-Baronniers in Dauphiné was wounded in the skirmish of Chasenay by three handgun shots
62 penetrating into his body, one of which he had beneath the larynx right near the pipe of the lung, passing near the nape of the neck, and the bullet is still there at present: as a result of which he has sustained several big and cruel accidents [side-effects], such as fever, a great tumor about the neck, so that he was unable to swallow anything for ten days, except some clear broth [from time to time], and despite all of these things he recovered his health and is at present still alive; and he was dressed by master Jacques Dalam, a very expert surgeon, living in the city of Montelimar in Dauphiné.

Alexander Benedict (Book 3, Chapter 5 of his *Histoire anatom.*) writes of a villager who was injured by an arrow-shot in the back, and it was extracted, but the iron [tip] remained in the body, which measured two fingerlengths across, and was barbed on the sides. The surgeon having looked for it for a long time without being able to find it, closed the wound, and two months later this iron [tip] came out, seemingly through the seat.

Moreover, in the same chapter he says that in Venice a girl swallowed a needle, which two years later she ejected while uri-

nating, covered with a stony matter, amassed around some sticky 63
fluid.

Similarly, Catherine Parlan, the wife of Guillaume Guerrier, a cloth merchant and a fine man, living on the rue de la Juisverie [i.e., the Jewish ghetto] was going to the fields mounted on a horse [when] a needle from her pin case entered into her right buttock in such a way that they couldn't pull it out of her. Said Parlan woman went two months without being able to remain seated, because she could feel the needle pricking her. Four months later she sent for me, complaining that when her husband made love to her she felt in the right groin a great stinging pain, because he was pressing upon it. Having put my hand on the [site of the] pain, I found an asperity and hardness, and I managed to extract from her said needle, completely rusted. This is surely to be placed in the rank of monstrous things, seeing that the steel, which is heavy, rose upward, and passed across the muscles of the thigh,
without causing an apostema. 64

15

OF STONES THAT ARE ENGENDERED IN THE HUMAN BODY

n the year 1566, the children of master Laurens
Collo—men very experienced in the extraction of 65
stones—took one out as thick as a walnut, in the middle of which was found a needle with which seamsters customarily sew. The patient's name was Pierre Cocquin, residing on the rue Gallande, near Maubert Square in Paris, and he is still living at the present writing. The stone was presented to the King in my presence, along with said needle, which the aforementioned Collos gave me to put in my study, which [needle/stone] I keep and still have in my possession at present as a token of such a monstrous thing.

In the year 1570, her highness the Duchess of Ferrara sent in this city for Jean Collo to come and extract a stone from the bladder of a poor pastry cook living in Montargis, which [stone] weighs nine ounces, [is] as thick as a fist, and of the form you see here in this picture; and it was extracted in the presence of milord master François Rousset and master Joseph Javelle, learned men and very experienced in medicine, physicians ordinary of the aforementioned lady. And it was so successfully extracted that said pastry cook was cured. (And said lady, accustomed to helping the poor, paid all the expenses for the cure of said pastry cook.) However, a little later on, he got a suppression of urine on account of two small stones which came down from his kidneys [and] which plugged the ureters (pores) and were the cause of his death.

In the year 1566 the brother of the aforementioned Jean Collo, named Laurens (said Collos, surgeons ordinary of the King, are very expert at the extraction of stones and at several other surgical operations) similarly performed in this city of Paris an extraction of three stones being in the bladder, each one of the thickness of

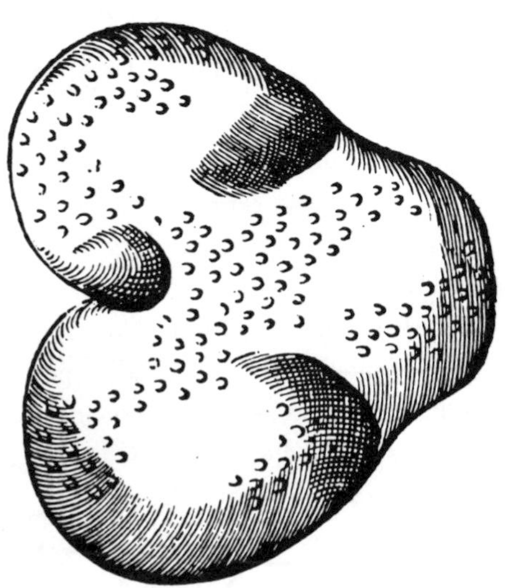

31. Figure of a stone extracted from a pastry cook of Montargis

a good fat hen's egg, white in color, the three of them weighing
twelve ounces and more, from a man (who, as soon as his wound
was closed up, returned to his house, where at present he still
lives). He was from Marly and his last name was Tire-vit [Pull-
cock], and because he had had some beginning of these stones in
his bladder since he was ten, he habitually pulled on his rod,
wherefore he was called "Tire-vit"; for the expulsive property [or
power] of the bladder, indeed of the whole body, was straining to
cast out what irritated it and, because of this, caused in him a
certain pricking sensation at the extremity of his rod (as occurs
ordinarily in those who have some sand or stone in the parts
connected with urinating, about which I have written more thoroughly in my book *Des Pierres*). (Those who have a stone in the
bladder always have an itching and stinging at the extremity of the
rod. [1573]) Those stones [from Tire-vit] were presented to the 66
king, being at the time at Saint-Maure-des-Fossés; they broke one
of them with a tapestry-maker's hammer, in the middle of which
was found another one, resembling a peach pit, tawny in color. The

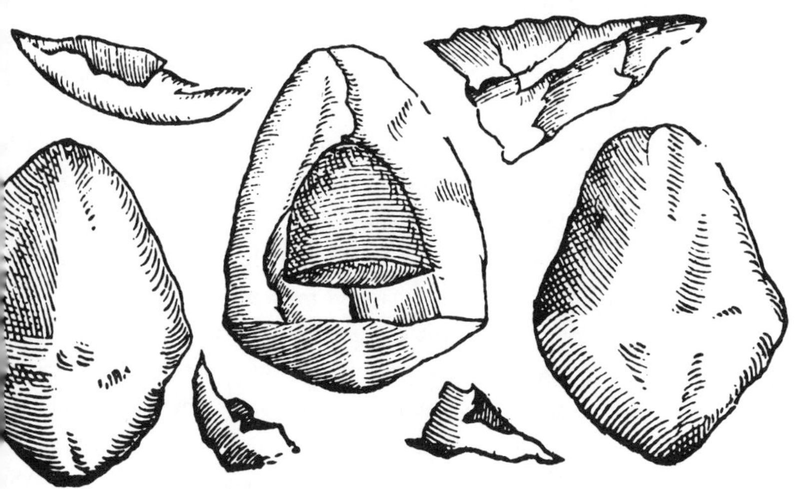

32. *Figures of three stones extracted at the same time without any time interval, from the bladder of a man called Tire-vit, one of which is broken*

aforementioned Collos gave me said stones to put in my cabinet as monstrous things, and I have had them depicted as close as
67 possible to life, just as you can see by these figures. [Fig. 32]

Moreover, I can here attest that I have found them in the kidneys of cadavers, in various shapes, such as of pigs, dogs, and other divese conformations, which has been left to us in writing by the Ancients.

Monsieur Dalechamps tells in his *Surgery* that he saw a man who had an apostema on the loins after the suppuration of which it degenerated into a fistula, through which he voided at various times several stones coming from the kidney; and yet he tolerated the stress of riding on a horse or in carriages.

Hippocrates (Book 5 of his *Epidemics*) writes of the chambermaid of Dyseris [in Larissa], sixty years of age, who had pains as if she were about to give birth [and] from whose womb a woman extracted a sharp hard stone, of the size, thickness, and conformation of a spindle balance [or whorl].

Jacques Hollier, regent Doctor on the faculty of Medicine at Paris writes (Book 1, Ch. on *Palpitation of the Heart*) that a woman having been tormented by difficulty in urinating over the space of four months, finally died; which [woman] being opened, there were found in the substance of the heart two rather large stones, with several little apostemas, the kidneys and ureters (pores) being healthy and intact.

In the year 1558 I was called by Jean Bourlier, master cloth tailor, living on the rue Saint Honoré to open for him a watery apostema that he had on his knee, in which [apostema] I found a stone as thick as an almond, very white, hard, and polished, and
68 he was cured and is still alive at the present writing.

A lady of our Court was extremely sick for a very long time feeling pain in the belly, with great straining sensations, being cared for by several physicians, who were ignorant of the source of this pain. They sent for me to find out if I could discover the cause
69 of her sickness. By the order of the physicians I examined her seat and her womb with the instruments proper to performing this [task] and for all that, I could not recognize her sickness. Monsieur Le

Grand ordered an enema for her and as she was getting rid of it she ejected a stone as thick as a thick walnut through the seat; and all of a sudden her pains and other symptoms ceased, and since then she has felt very good. 70

A similar thing happened to a lady from Saint Eustache, who lived at the intersection of the rue de la Harpe.

Captain Augustin, Engineer of the King, sent for me together with Monsieur Violaine, regent doctor of the faculty of Medicine and Claude Viard, Sworn Surgeon in Paris, to extract from him a stone that he had under his tongue, a half-a-finger long and as thick as a feather quill. He has another one of them that cannot 71
yet be readily detached.

Now to say it in a word, stones can be engendered in all parts of the body, interior as well as exterior. That it be true, one can observe them engendered in the joints of gouty persons. Antonius Benivenius, a Florentine physician (in Book 1, chap. 24) says that a man named Henry Alleman ejected a stone of the thickness of a filbert while he was coughing.

16
ON CERTAIN MONSTROUS ANIMALS THAT ARE BORN ABNORMALLY IN THE BODIES OF MEN, WOMEN, AND SMALL CHILDREN

Just as in the big world [i.e., the macrocosm] there are two great lights, to wit, the sun and the moon, so there are in the human body two eyes which illuminate it, which [body] is called the microcosm or small portrait of the big world, abridged. Which [microcosm] is composed of 72
four elements, as is the big world in which winds, thunder, earthquakes, rain, dew, vapors, exhalations, hail, eclipses, floods, sterility,

fertility, stones, mountains, fruits, and several diverse species of animals occur; the same thing also happens in the small world, which is the human body. An example of winds: they can be observed to be enclosed in windy apostemas and in the bowels of those who have windy colic; and similarly in some women whose belly one can hear rumbling in such a way that it seems there is a colony of frogs there; the which [winds] upon issuing from the seat make noises like cannons being fired. And although the [artillery] piece is aimed toward the ground, nevertheless the cannon smoke always hits the nose of the cannoneer and those who are near him.

An example of rains and floods: this may be seen in watery apostemas and in the belly of those afflicted with dropsy. An example of an earthquake: such a thing can be seen at the onset of an attack of ague, in which the poor persons sick with fever have a shaking all over in their body. An example of eclipse: this can be seen in syncopes [i.e., faintings] or in heart failure and in suffocations of the womb. An example of stones: one sees them in those persons from whose bladder or other parts of the body one has extracted them.

An example of fruits: how many [persons] does one see who on their face or other parts of the body have the form of a cherry, of a plum, of a sorb-apple, of a fig, or a mulberry, the cause of which has always been referred to the very powerful imagination of the conceiving or pregnant woman moved by a vehement appetite or with the appearance of an unexpected touching of this [fruit]; so likewise because one sees some persons born from it [such imagination?] having in some part of the body the shape and substance of bacon rind, others of a mouse, others of a crawfish, others of a sole, and other such; which is not beyond reason, given the force of the imagination being joined with the conformational power, the softness of the embryo, ready like soft wax to receive any form, and given that when one might seek to examine all those who are thus marked, it will be found that their mothers have been moved during their pregnancy by some such appetite or happening. Whereupon we will note in passing how dangerous it is to disturb a pregnant woman, to show her or to remind her of some food which she cannot enjoy immediately, and indeed to show them animals, or even pictures of them, when they are deformed and monstrous. For which I'm expecting someone to object to me that I therefore shouldn't have inserted anything like this into my book on reproduction. But I will answer him in a word, that I do not write for women at all. Let us return to our subject.

An example of mountains: one sees them on hunchbacks and on those who have thick and enormous bunches [or growths]. An example of sterility and drought: one sees it in those suffering from a hectic [constant consuming fever] who have the flesh of their body almost totally consumed. An example of fertility: one recognizes it in those who are fat with large buttocks[es] and paunches, so much so that they are bursting in their skin, [and] it is necessary for them to remain always lying down or seated, because they cannot carry the great mass of their body. An example of animals that procreate on and in our bodies, to wit, lice, bedbugs, and crabs, and others that we shall presently describe.

Monsieur Houlier writes in his *Pratique* that he treated an Italian tormented with an extreme headache, from which he died. And having had him opened, there was found, inside the substance of his brain, a scorpion which as the said Houlier thinks, had been engendered in him for having continually sniffed basil. Which is very likely, in view of the fact that Chrysippus, Diphanes, and Pliny have written that if basil is crushed between two stones and exposed to the sun a scorpion will be born of it.

Monsieur Fernel writes of a soldier who was very flat-nosed, so much so that he could not blow his nose at all, as a result of which two hairy and horned worms a half-finger thick were born from the retained and rotten excrement, which [worms] drove him mad in the space of twenty days and were the cause of his death. 73

Not long ago a young man had an apostema in the middle of the external part of the thigh, from which this animal issued, [and] which was brought to me by Jacques Guillemeau, Surgeon Ordinary of the King, who said he had extracted it: and I placed it in a glass vial and it remained alive for more than a month without any food. The figure is represented to you here:

33. [Figure of an animal in an apostema (ed. Malgaigne)]

34. [Figure of an animallike material ejected by Count Charles de Mansfeld (ed. Malgaigne)]

Monsieur Duret assures me that he ejected through his rod, after a long illness, a live animal similar to a woodlouse [or centipede] which the Italians call *Porceleti,* which was red in color.

Monsieur the Count Charles de Mansfeld, recently being sick at the hôtel de Guise with a high and constant fever, ejected through his rod a certain matter similar to an animal, a figure of which is
74 reproduced for you here.

Many animal forms are likewise created in women's wombs (which are often found with fetuses and well-formed young), such as frogs, toads, snakes, lizards, and harpies.

(Women have been observed casting out of their wombs snakes and other creatures, which can happen through the corruption of certain excrements being retained in their womb, as one sees occur in the intestines and other parts of our bodies, thick long worms, indeed hairy and horned, as we shall show hereafter. Some people have tried to claim that such a thing can come about when a woman bathes, if by chance some poisonous creature such as snakes and others have spawned and cast their seeds in the water, {or} at the spot where it may have happened that someone has drawn such an ordure up from the well along with the water and that then afterward the woman has bathed in it, seeing principally that because of sweat and heat, all her pores are the more open. But such a cause cannot take place, given that the generative force of this seed is suffocated and extinguished by the great quantity of hot water, including the fact, also, that the mouth of the womb does not open at all except at the time of coitus, or when the monthlies are flowing. {Ch. 19, *Des Monstres,* 1573})

Nicole Florentin compares {these animals or harpies} to horn owls and says they should be called wild beasts. The ancients called harpies Lombard brothers because such things happened to the women of Lombardy and because they were born in one and the same womb like well-formed young, which gave occasion for calling them "uterine brothers," to say something nasty about a person one hates (Gourdon, Book 7, Chapter 18). Now the women of the realm of Naples are very subject to this because of the bad food they eat, for they have from time immemorial preferred to have
a muslin stomach to a velvet one, that is to say, {they prefer} to 75
eat fruit, grasses {and/or herbs} and other bad-tasting and unnutritious things, which generate such animals through putrefaction, than to eat good nourishing food, just in order to be sparing and to be elegant and trim.

Monsieur Joubert (in his book *On Errors Made by the People*) 76
writes of two Italian ladies: one was the wife of a second-hand {clothes} dealer, the other {a velvet-hooded} gentlewoman, {but} each one of them delivered a monstrous birth within the same month; that of the clothes merchant's wife was small, resembling a tailless rat, the other of the gentlewoman's was fat, like a cat; they were black in color, and upon leaving their wombs these monsters climbed up to the space between the wall and the bed and attached themselves firmly to the bedpost.

Lycosthenes (in his prodiges) writes that in the year 1494 a woman of Krakow, [living] on the square called Holy Spirit, gave birth to a dead child who had a live snake attached to its back, who was gnawing on this little dead creature (as you see in this
77 picture).

Levinus (Book I, chap. 8, *De occult. natur.*) tells a story we must marvel at, [and he tells it] in this fashion: These few years past a woman came to me to ask my advice, who, having conceived by a mariner, her womb began to swell in such a way that they thought she was never going to carry [the child] to term. The ninth month having passed by, she sends for the midwife, and with great effort first she delivered a formless mass of flesh, having on each side handles the length of an arm, which moved and had life, like sponges. Afterward there came out of her womb a monster having a hooked nose, a long neck, sparkling eyes, a pointed tail, and very active feet. As soon as said monster had issued forth, he began to hum loudly and to fill the whole chamber with whistlings, running here and there to hide himself, upon which [monster] the women threw themselves and suffocated him with pillows. In the end, the poor woman, completely exhausted and torn, gave birth to a male child so racked and tormented by this monster that he died as soon as he had received baptism. The aforementioned patient, after

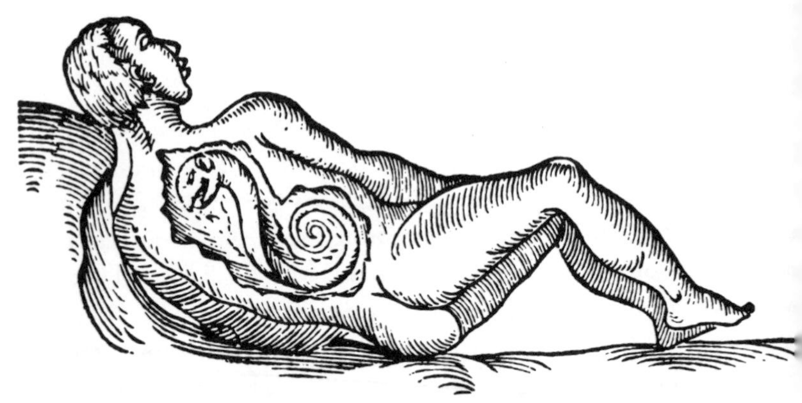

35. Figure of a child who had a live snake in its back

having taken a long space of time to recover, told him the whole thing faithfully. 78

Cornelius Gemma, a doctor from Louvain, in a book he wrote a short while ago, entitled *De naturae divinis characterismis,* recounts a story we must wonder at about a girl in said town, fifteen years of age, from whose body top and bottom, issued after endless pain several strange things, among which she cast out, through the seat, along with excrement, a live animal a foot and a half long, thicker than the thumb, representing a real, natural eel so much that there was nothing to say it wasn't, except that it had a very hairy tail (as you can see by the picture below [omitted], similar to the one that Gemma put in his book). 79

Master Pierre Barque, a surgeon in the French troops, and Claude Le Grand, a surgeon, both living in Verdun, not long ago assured me that they had cared for the wife of a man named Grasbonnet, living in the aforementioned Verdun, and this wife had an apostema in the belly; and when it was opened there issued forth along with the pus a great number of worms, as thick as a finger, having pointed heads, and these worms had gnawed her intestines, as a result of which for a long time she evacuated her fecal excrement through the ulcer, and she is at present completely cured.

Antonius Benivenius, a physician of Florence, writes that a certain fellow named Jean, a joiner, forty years of age, had an almost 80
constant heart pain, on account of which he had been in danger of death. And in order to get rid of it he got the opinion of several physicians of his time, without, even so, having received any relief from them. Some time afterward he approached him [Benivenius]: having considered his pain, he gave him a vomitive, by which he ejected a great quantity of rotten and corrupt matter, without, even so, alleviating his pain. He immediately ordered another vomitive for him, by means of which he vomited a great quantity of matter, together with a worm four fingers long, the head red [and] round, and as thick as a fat pea, having its body full of down and its tail forked in the form of a crescent, together with four feet, two before and two behind.

I say again that in apostemas one finds very strange bodies, such as stones, chalk, gravel, coal, snail shells, blades of wheat or grass hay, horns, hair, and other things, together with several and diverse
81 animals, both living and dead. The generation of these things (accomplished through corruption and diverse alteration) should not surprise us much, if we consider that, since fecund Nature has proportionately in the excellent microcosm every sort of matter in order to make it resemble and be like a living image of the big world [the macrocosm], so she struggles to represent in it [i.e., the microcosm] all her actions and movements, never being idle, so long as she is not lacking in matter.

(I have written in my *Treatise on the Plague* [1568] about having seen a woman who had ejected a worm through the seat, the length of a fathom [ca. 6 feet], having the shape of a snake; he who wishes to know about the generation, species and difference, their various colors, and the shapes of these will find them in said
82 chapter. [1573]).

17

ON CERTAIN STRANGE THINGS THAT NATURE REPELS THROUGH HER INFINITE PROVIDENCE

ntonius Benivenius, a physician of Florence writes that a certain female swallowed a brass needle without having felt any pain for the space of a year which [time] having passed, a great pain in the stomach came upon her, and for this [reason] she got the opinion of several physicians concerning this pain, without mentioning to them this needle which she had swallowed: however, not one [of them] was able to give
83 her relief, and she lived in that condition for the space of ten years then all of a sudden said needle came out through a little hole near
84 her navel, and she was cured in a short while.

A schoolboy named Chambellant, a native of Bourges studying
n Paris at the Collège de Presle, swallowed a blade of grass called
ramen, which some time afterward issued intact from between his 85
ibs, from which he thought he would die, and he was cared for
y the now defunct Monsieur Fernel and Monsieur Huguet, Doc-
ors on the faculty of Medicine. It seems to me it was quite a thing
or Nature to have expelled said blade from the lung substance,
o have made an opening in the pleuritic membrane and in the 86
nuscles that are between the ribs, and nevertheless he was healed,
nd I think that he is still living.

Cabrolle, the surgeon of monsieur le Mareschal d'Anville, not 87
ong ago assured me that François Guillement, a surgeon of Som-
nières, a small city which is about four leagues from Montpellier,
ad cared for and cured a shepherd whom some thieves had forced
o swallow a knife half a foot long, and the handle was of horn,
s thick as one's thumb; which was for the space of six months
n his body [while he was] complaining enormously; and he became
ectic [wasted], dried up and emaciated; finally he got an apostema
eneath the groin, ejecting a great quantity of very bad smelling
nd infected pus, through which was extracted said knife, which
Monsieur Joubert, a famous physician at Montpellier, keeps in his
tudy and he has shown it to several [persons], as a thing to be
vondered at and worthy of remembering, and monstrous. Which
ikewise Jacques Guillemeau, a Sworn Surgeon in Paris, affirmed
o me that he had seen in Monsieur Joubert's study, then being
n Montpellier.

Monsieur de Rohan had a fool named Guion who swallowed the
ip of a sharp sword, three fingers long, or thereabouts, and twelve
ays later he ejected it through the seat, and this didn't occur
vithout a lot of side effects, yet he escaped; there are some gentle-
nen from Brittany still living who saw him swallow it.

It has been observed also in certain women, their child being
ead in their womb, [that] the bones come out of their umbilicus
nd the flesh through putrefaction is ejected through the neck of
heir womb and through the seat, having created an abcess: which

two famous surgeons, worthy of our believing, assured me of havin seen in two different women.

Likewise monsieur Dalechamps in his *Chirurgie françoise* relate that Albucrasis had treated a lady of the same thing, the outcom of which was good, having recovered her health, without, howeve carrying any children since then.

Similarly, it is a monstrous thing to see a woman—on accour of a suffocation [or fit] of the womb—be three days without moving without appearing to breathe, without any apparent arterial pul sation; on account of which some women have been buried aliv their friends thinking that they were dead.

Monsieur Fernel writes of a certain adolescent, who after havin taken a good deal of exercise, began to cough until he had ejecte a whole apostema as thick as an egg, which being opened, wa found [to be] full of white slime enveloped in a membrane. Th youth, having spit up blood for two days, [and] with a high feve nevertheless escaped.

The child of a cloth merchant named de-Pleurs, living at th corner of the rue neuve Notre-Dame-de-Paris, twenty-two month of age, swallowed a piece of steel mirror, which descended into th

36. Figure of a piece of mirror, which a child twenty-two months of age swallowed, which was the cause of his death

ag [scrotum] and was the cause of his death. Being deceased, he
vas opened in the presence of monsieur le Gros, regent doctor on
he faculty of Medicine in Paris, and the opening was executed by
naster Balthazar, then surgeon at the Hôtel-Dieu. Curious about 88
he truth, I went to speak to the wife of said de-Pleurs, who assured
ne the thing was true and showed me the piece of mirror that she
arried in her purse; which was of such formation and size.

Valescus de Tarante, physician, in his *Medical Observations and
are examples,* says that a young Venetian girl swallowed a needle 89
hile sleeping, four fingers long, and ten months later she ejected
through the bladder along with urine.

In the year 1578 in the month of October, Tiennette Chartier,
ving in Saint-Maur-les-Fossés, a widowed woman forty years of
ge, being sick from a tertian ague [three-day fever], vomited at
ne beginning of her attack a great quantity of bilious juice, with
hich she cast up three worms, which were hairy and in every
espect identical—in form, color, length, and thickness—to cat-
rpillars, except that they were blacker; which [worms] lived eight
ays and more without any food. And these were taken by the
arber of said Saint-Maur to monsieur Milot, doctor and reader
f the schools of Medicine, who then cared for the aforementioned
hartier woman [and] who showed them to me. Messieurs le Fèvre,
Gros, Marescot, and Courtin, Doctors of Medicine, also saw
nem there.

I cannot yet conclude without relating this story taken from the
hroniques of Monstrelet, about a *franc-archer* of Meudon, near 90
aris, who was a prisoner at Châtelet for several larcenies for which
e was condemned to be hanged and strangled; he appealed to the
ourt of Parliament, and by that court he was declared to have
een correctly judged and to have had no cause for appeal. On
e same day the King was warned by the physicians of that city
at several [persons] were in great travail and affliction from stone,
olic, suffering, and sickness in the side, with which said *franc-
cher* was [also] afflicted; and [that] monseigneur de Boscage was
so afflicted with said malady, and [the king was told] that it would
very desirable to see the places where said illnesses are created

within human bodies, which thing could not be better known tha by cutting into the body of a living man: which could very well b done on the person of this *franc-archer,* who was also very close t suffering death; which opening was executed on the body of sai *franc-archer,* and within this [man] the place of such illnesses wa looked at and examined, and after they had finished looking he wa sewn up again, and his entrails put back; and through an orde from the King, he was bandaged and cared for well, so much s that he was completely cured within a few days; and he got hi remission [pardon] and money was given to him along with it.

18

AN EXAMPLE OF SEVERAL OTHER STRANGE THINGS

lexander Benedict writes in his *Pratique* of hav ing seen a woman named Victoire who had lost all h teeth, and having become bald, another set of tee came back to her at the age of eighty.

Antonius Benivenius, a physician—in Book 1, Chap. 83—me tions a man named Jacques the thief, who having deceased, h heart was found to be completely covered with body hair.

The son of Bermon, a *Baille* [a city administrator] living in t city of St.-Didier, in the country of Vellay, had a growth on t brow of his right eye which was already beginning to obfuscate a cover it, and therefore he wanted me to amputate it (which I d not long ago) and I found the growth full of body hair [togethe with [a] mucilaginous matter; and in eight days the wound w
91 completely closed up.

Estienne Tessier, a master barber-surgeon living in Orleans, very honest man and experienced in his art, related to me that short while ago he had cared for and doctored Charles Verign a sergeant living in Orleans, for a wound he had received in the ham, the right part, with total incision of the two tendons which flex the ham, and, in order to dress it, he had him flex his leg, so that he sewed the tendons end to end one to the other, and he set it and treated it so well that the wound was closed up without [the patient's] having been left a cripple: a thing worthy of being noted by any young surgeon, so that when such a thing comes into his hands, he may do likewise.

What more shall I say? That I've seen several [persons] cured, having sword, arrow, and handgun wounds across their bodies; others from head wounds, with loss of brain substance; others have arms and legs carried off by cannon fire, [and] nonetheless be healed; and others who had only small, superficial wounds—that one estimated to be nothing—nonetheless die with great and cruel side effects. Hippocrates (in the fifth [book] of his [work] *On Epidemics*) says he has torn out six years after [it had penetrated] and arrowhead which had remained in the deepest [part] of the groin and [he] gives no other cause for its remaining so long, except that it had lodged among the nerves, veins, and arteries without wounding a single one [of them]. And by way of conclusion, I shall say along with Hippocrates (father and author of Medicine) that in illnesses there is something divine, which man cannot rationalize [or explain]. I would mention here several other monstrous things that occur in illnesses, if it weren't that I fear being prolix, and repeating a thing too many times.

19

AN EXAMPLE OF MONSTERS CREATED THROUGH CORRUPTION AND PUTREFACTION

oistuau, in his *Histoires prodigieuses,* writes that when he was in Avignon, an artisan, while opening a deadman's lead casket having a very good cover and being well soldered—so that there was no air in it—was bitten by a snake which was enclosed within, whose bite was so poisonous that he thought he would die from it. One can easily explain the birth and the life of this animal; it is because it was engendered in the putrefaction of the dead body.

92 Baptiste Leon writes similarly that in the time of Pope Martin V, there was found enclosed in a large solid stone a live snake having no apparent mark [of hole] through which he might have breathed.

At this spot I want to relate a similar story. Being in one of my
93 vineyards near the town of Meudon, where I was having some big, thick, solid stones split, they found in the middle of one of them a fat live toad and [yet] there was no sign of an opening in it, and I wondered at how this animal had been able to be born, to grow, and to be alive. Then the quarrier told me that one should not wonder at it, because several times he had found such, and other animals [too], at the deep [center] of stones that had no sign of any opening. One can also explain the birth and life of these animals: it is that they are engendered with some humid substance from the stones, the which putrefied humidity produces such creatures.

20

AN EXAMPLE OF THE MIXTURE OR MINGLING OF SEED

here are monsters that are born with a form that is half-animal and the other [half] human, or retaining everything [about them] from animals, which are produced by sodomists and atheists who "join together" and 94
break out of their bounds—unnaturally—with animals, and from this are born several hideous monsters that bring great shame to those who look at them or speak of them. Yet, the dishonesty lies in the deed and not in words; and it is, when it is done, a very unfortunate and abominable thing, and a great horror for a man or a woman to mix with or copulate with brute animals; and as a result, some are born half-men and half-animals.

The same occurs if animals of diverse species cohabit with one another, because Nature always strives to recreate its likeness; thus there has been seen a lamb having the head of a pig, because a boar had covered the ewe; for we see even in inanimate things, such as wheat coming from a grain of wheat—and not barley—and an apricot tree coming from an apricot pit—and not the apple tree—how Nature always preserves its kind and species.

In the year 1493 a child was conceived and engendered of a woman and of a dog, having, from the navel up, upper parts similar in form and shape to the mother, and it was very complete, without Nature's having omitted anything; and, from the navel down, all its lower parts were also similar in form and shape to the animal, that was the father: which (just as Volateranus writes) was sent to the pope who reigned at that time. Cardan, book 14. Chap. 64, *On the variety of things* mentions it. [Fig. 37] 95

Coelius Rhodiginus, in his *Readings from the Ancients* (Lib. 25, ch. 32 [1573]), says that a shepherd named Cratain in Sybaris 96
having exercised his brutal desire with one of his goats, the goat, some time afterward, kidded a kid which had a head of human form

and similar to the shepherd but the rest of the body resembled the goat.

37. Figure of a child, part dog

In the year 1110 a sow in a burg near Liège farrowed a pig having the head and face of a man, as were the hands and feet,
97 but the rest [was] like a pig. [Fig. 39]

In the year 1564 in Brussels, at the house of a man named Joest Dickpert, living on Warmoesbroeck street, a sow farrowed six pigs,

38. Figure of a monster with the face of a man and the body of a goat

the first of which was a monster having a man's face as well as arms and hands, representing humanity in a general way from the shoulders [up]; but the two hind legs and hindquarters of a swine, [and] having the *nature* [genitals] of a sow: it nursed like the others and lived two days: then it was killed along with the sow on account of the horror that people had of it (whose portrait is shown to you here, as realistically as possible). 98

In the year 1571 at Antwerp, the wife of a companion printer named Michel living at the home of Jean Mollin, a sculptor at the sign of the Golden Foot, on Camerstrate, on a proper [feast] day of Saint Thomas, about ten o'clock in the morning, gave birth to a monster representing the shape of a real dog, except that he had a very short neck, and the head more or less of a fowl, except without hair; and it had no life at all, because said woman delivered before term; and at the very hour of her delivery, uttering a horrible cry (a thing to be marvelled at) the fireplace of the house fell to earth without in the least bit harming four little children who were

39. Figure of a pig, having the head, hands, and feet of a man and the rest of a pig

40. A monster half-man, half-swine

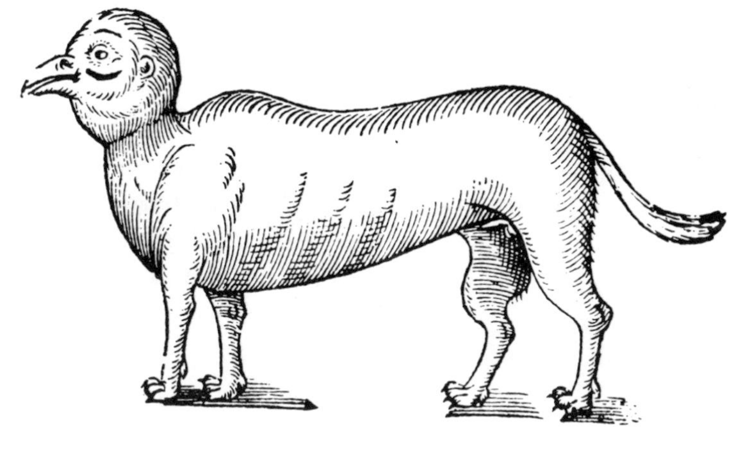

41. Figure of a monster like a dog, with a head like a bird

gathered around the hearth. (And because it is a recent thing, it 99
seemed to me a good thing *for posterity* to give a portrait of it
here). [Fig. 41] 100

(In the year 1224 near Verona, a mare foaled a colt which had a well-formed man's head, and the rest [of it] was [the body of] a horse. The monster had a man's voice, at whose cry a villager of the region came running, and being astonished to see so horrible a monster, he killed it; for this reason, being taken to court and interrogated as much regarding the birth of this monster as regarding the reason why he had killed it, he said that the horror and the fear that he had had of it had made him do it, and as a result he was absolved. [1573, 1575])

Loys Celée writes of having read in a respected author that a ewe conceived and lambed a lion, a monstrous thing in nature.

On the thirteenth day of April 1573 a lamb was born at a place called Chambenoist on the outskirts of Sezanne, in the house of
Jean Poulet, a salt surveyor, and there was no life known to this 101

lamb except that it was seen to move just a very little bit; beneath its ear was an opening approximating the shape of a lamprey.

This present year, 1577, a lamb was born in the village named
Blandy, a league and a half from Melun, having three heads in one,
and when one of said heads bleated, the others did the same. Master
Jean Bellanger, a surgeon living in the city of Melun, affirmed that
he had seen it, and he had the figure of it portrayed, which was
102 barked and sold through this city of Paris with *privilège,* together
with [pictures of] two other monsters, one of two twin girls, and
another having the face of a frog, which has been shown in a
previous illustration.

42. The figure of a lamb having three heads

There are divine things, hidden, and to be wondered at, in monsters—principally in those that occur completely against nature [i.e., unnaturally]; for in those, philosophical principles are at a want, so that one cannot give any definite opinion in their case. Aristotle in his *Problems* says that some monsters are formed in nature [i.e., naturally], because of the badly disposed condition of the womb and [because of] the course of certain constellations. Which happened in the time of Albert [Albertus Magnus], on a farm, when a cow had a calf [that was] half-man: for which the villagers—suspecting the cowherder—accused him and sentenced him, intending to have him burned with said cow; but Albert, on account of having done several experiments in astrology, knew (said he) the truth of the event, and he said that that had come about through a special constellation, with the result that the cowherder was freed and purged of the imposition of such a crime. I strongly doubt whether the judgment of Lord Albert was good (because God is not tied nor subject to following the order He has established in Nature, nor in the movement of stars and planets. The opinion of astrologers is very suspect, and I leave it to them to dispute and
prove. Jeremiah, 10. God is not at all subject to the 103
stars, for He is the author of all things. Book of the Ephesians.
[1573, 1575]). 104

Now I shall refrain from writing here about several other monsters engendered from such grist, together with their portraits, which are so hideous and abominable, not only to see but also to hear tell of, that, due to their great loathsomeness I have neither wanted to relate them nor to have them portrayed. For (as Boistuau says, after having related several sacred and profane stories, which are all filled with grievous punishments for lechers) what can atheists and sodomists expect, who (as I said above)
couple against God and Nature with brute animals? On this 105
subject, Saint Augustine says the punishment of lechers is to fall into blindness and to become insane, after they have forsaken God, and not to see [i.e., be aware of] their blindness, being unable to follow good counsel [and] provoking the wrath of God against them.

21

AN EXAMPLE OF THE ARTIFICE OF WICKED SPITAL BEGGARS

106 I have a recollection, being in Angers, in 1525, that a wicked scoundrel {and beggar} had cut off a hanged man's arm—already stinking and infected—which he had tied to his vest, letting it lean on a small fork against his side, and he hid his natural arm behind his back, covered with his cloak, so that people would think that the hanged man's arm
107 was his own; and he shouted at the temple {Protestant church?} door that people should give him alms, for Saint Anthony's sake. One Good Friday, the people, seeing his rotted arm, gave him alms, thinking it was real. The beggar having wiggled this arm around for a long time, finally it came loose and fell to the ground, where as he was quickly trying to pick it up, he was perceived by some to have two good arms, not counting that of the hanged man; thereupon he was led off as a prisoner, then condemned to get the whip, by order of the magistrate, with the rotted arm hung around his neck, in front of his stomach, and {to be} banished forever from the country.

22

THE IMPOSTURE OF A WOMAN BEGGAR WHO PRETENDED TO HAVE A CANKER ON HER BREAST

A brother of mine named Jehan Paré, a surgeon living in Vitré, a city in Brittany, saw a fat, plump hedge-whore and beggar begging alms at the door of a temple {Protestant church?} one Sunday who pretended to have a canker {cancerous ulcer} on her breast, which was a very hideous thing to see, because of a great amount of foul matter which seemed to flow out onto a linen cloth that she had on her front. My afore-

mentioned brother, studying her face, which was of a good, lively color, showing her to be very healthy, and [observing] the areas around her ulcerated canker [to be] white and of good color and the rest of her body [to be in] very good condition, he thought to himself that this wench could not have a canker, being so fat, plump, and full-cheeked, and he was sure that it was an imposture; the which he denounced to the magistrate (in that country called *l'Aloué* [a Seneschal's lieutenant]) who permitted my aforementioned brother to have her brought to his place in order to know more certainly that it was an imposture. Which [woman] having arrived there, he uncovered her whole chest and found that she had, under her armpit, a sponge soaked and imbued with animal blood and with milk mixed together, and a little elder pipe through which this mixture was conducted through fake holes in her ulcerated canker, flowing onto the linen that she had on her front: and by that he knew for certain that the canker was artificial. Then he took some warm water and fomented the breast, and having moistened it, he lifted off several black, green, and yellowish frogskins, placed one on top of the other, stuck together with bolearmenie and egg white 108
and flour, which they found out through her confession, and having lifted all of them off, they found the teat healthy and whole and in as good condition as the other. Once this imposture was discovered, said *Aloué* had her declared prisoner, and she, being interrogated, confessed the imposture and said that it was her bedmate [also a beggar], who had rigged her up like that, who [himself] seemingly was pretending to have a great ulcer on his leg, which gave the semblance of being real by means of an ox's spleen that he placed along and around his leg, attached and pierced very aptly, with old rags at the two extremities, so that it seemed to be two times bigger than normal; and to make the thing more monstrous and hideous to see, he made several cavities in said spleen and on top he spread this mixture made of blood and milk and over all his rags. The aforementioned *Aloué* sent for this master-beggar, thief, imposture, who could not be found; and he condemned the slut to get the whip, and [she was] banished out of the country, but not before being curried with lashes of a whip made of knotted cords, as one did at that time.

23

THE IMPOSTURE OF A CERTAIN BEGGAR WHO WAS COUNTERFEITING A LEPER

year afterward there came a big knave of a beggar who counterfeited a leper; he placed himself at the door of the [Protestant?] church, unfolding his wares, which included a kerchief on which he placed his money barrel and several kinds of coins, holding in his right hand some
109 clappers, making them click rather loudly: his face covered with large pustules, made of a certain strong glue and painted in a livid, reddish fashion, approximating the color of lepers, and he was very hideous to see; thus through [their] compassion everyone gave him alms. My aforementioned brother went up to him and asked him how long he had been sick like that; he answered in a broken and rusty voice that he had been a leper from his mother's womb and that his father and mother had died of it, and that their limbs had fallen off of them piecemeal because of it. This leper had a certain strip of cloth twisted around his neck and beneath his cloak, with his left hand, he squeezed his throat, so as to make the blood mount to his face, in order to render it still more hideous and disfigured, and also in order to make his voice husky, which came about through the distress and stricture of the windpipe squeezed by the strip of cloth. With my aforementioned brother staying there to chat with him thus, the "leper" couldn't go for such a long time without letting loose of the strip of cloth so that he could get his wind back a bit; which my aforementioned brother noticed, and it is thus that he had a suspicion that there was some falseness and imposture [at stake]. For this reason he went directly to the Magistrate, asking him to be so good as to help him find out the truth about it; which he gladly granted him, ordering that he [the "leper"] be brought to his house to investigate as to whether he was leprous. The first thing that he [my brother] did was to take the swathe that was around his neck off; then he washed his face with warm water which caused all his pustules to become detached and to fall

off; and his face remained full of life and natural, without any sign of disease. That done, he had him strip naked, and he found on
his body no sign of leprosy, either univocal or equivocal. The 110
magistrate being informed of this, he had him declared prisoner, and three days afterward he was interrogated; whereupon he confessed the truth (which he could not deny) after a long remonstrance that the Magistrate gave him, pointing out to him that he was a thief of the people, being healthy and sound enough to work. This "leper" told him that he didn't know any other trade other than to counterfeit those who are afflicted with St. John's disease, or
that of St. Fiacre, or of St. Main, in short, that he knew how to 111
counterfeit several illnesses, and that he had never found greater profit in it than when he counterfeited the leper: then he was condemned to get the whip on three successive Saturdays, with his money barrel hung around his neck on top of his chest, and his clappers over his back, and [to be] banished forever from the country, on pain of the halter [i.e., being hanged]. When it came to the last Saturday, the people shouted at the top of their voices to the executioner: "Strike, strike, officer! He can't feel anything; he's a leper!" wherefore at the voice of the people the executioner was cruelly bent on whipping him so hard that shortly afterward he died, both because of the last whipping and because of having opened up his wounds again three different times: a thing which didn't amount to any great loss for the country.

Some [of these beggars] ask to be lodged and to be fed at night; and once one has, through pity, taken them in, they open the doors and let their companions in, which latter pillage and often kill those who have put them up; thus an honest man, in good faith, often will be killed and pillaged by such wicked men, which has been seen frequently.

Others wrap their heads up in some miserable rag and sleep in the dung in certain spots where people are passing, begging alms with a low and trembling voice, like those who are coming down with an ague, and thus while they are pretending to be very ill, the people, taking pity on them, give them money; and yet they have no disease.

They have a certain slang by which they recognize and understand each other, so as to deceive people better, and under the cloak of compassion, people give them alms, which [only] supports them in their wickedness and imposture.

Women pretend to be pregnant, indeed on the point of delivery, putting a feather pillow over their belly, and they beg for linen and other things necessary for their confinement; which just a while ago I uncovered in this city of Paris.

Others say that they are sick with icterus and that they have the jaundice, besmearing their whole face, their arms, legs, and chest with soot [of incense: bleach?] thinned out with water: but such an imposture is easy to uncover, just by looking at the whites of their eyes; for that is the part of the body in which said jaundice reveals itself first; or else, if one rubs their face with a piece of linen soaked in water, their deceit is uncovered. To be sure, such thieves, malingerers and impostures, in order to live in idleness, do not ever want to learn any other skill than such beggarliness, which in truth is a study in thoroughgoing wickedness: for what persons could one find more apt at exercising the pimp and bawd trade, at strewing poisons throughout the towns and cities, committing treasons and serving as spies, at stealing, at highway robbery, and every other wickedness that one might practice? For over and above those who have bruised themselves and who have cauterized and stigmatized their bodies or who have used herbs and drugs to make wounds and bodies more hideous, there have [even] been some who have kidnapped little children and have broken their arms and legs, poked out their eyes, cut off their tongues, pressed upon and caved in their chests, saying that lightning bruised them thus when (carrying them about among people) they themselves have done this to give them the appearance of beggars and to get a few pennies.

Others take two children and put them in two baskets on top of an ass, shouting that they have been despoiled and their house burned. Others take a sheep paunch, fitting it over the lower part of their bellies, saying they've been ruptured and injured and that it is necessary to operate on them and amputate their testicles.

Others make their way on two little blocks who can tumble and turn somersaults as well as a gymnast. Others feign that they come from Jerusalem and that they are bringing back little trinkets for relics; and they sell them to the good village people. Others have a leg hung around their neck; others counterfeit being blind, deaf, impotent, making their way on two crutches, but otherwise they're good companions! 112

What more shall I say? That they divide up the provinces [among them] in order to bring back—after a certain time—bounty for their common good, [while] pretending to take a trip to Saint Claude, Saint Main, Saint Mathurin, Saint Hubert, or Our Lady of Loretto, in Jerusalem, and they are sent thus to see the world and learn; by whom they send [word] from city to city to their beggar friends—in their [private] lingo [as to] what they know that's new and that might concern their trade—such as some newly invented way of behaving that will get money.

Then not long ago a fat beggar pretended to be deaf, mute, and crippled; yet by means of a silver instrument he said he'd gotten in Barbary (nonetheless stamped with the stamp of Paris) he spoke so that one could understand him. He was discerned to be an imposter and was placed in the Saint Benedict Prisons and at the request of the Baillif [or Magistrate] of the Poor, I went to said prisons to examine said beggar in company [with other surgeons], and we made a report to the gentlemen at the Bureau of the Poor of Paris, as follows:

> We, Ambroise Paré, Counsellor and Chief Surgeon of the King; Pierre Pigray, Surgeon ordinary of his Majesty, and Claude Viard, Surgeon in Paris, do certify that on this very day, at the request of the Procurer of the poor, we have seen and examined in the Saint Benedict Prisons an individual who refused to tell his name, forty years of age, or thereabouts on whom we have found a third of the right ear lost, which was cut off from him. Similarly a mark on his right shoulder which we deem to have been made by a hot iron. Moreover, he was counterfeiting a

great shaking of his leg, this fellow saying it came from a loss of the hip bone, which is a false thing, all the more because said bone is there in [its]entirety; and there appears to be no sign by which we might say that the shaking comes from any sickness which might have gone before, but [rather that it] comes from a voluntary movement. *Item,* we inspected his mouth (due to the fact that he wanted to persuade us that his tongue had been pulled out of him by the nape of the neck, an enormous imposture and which cannot be done), but we found his tongue intact without any lesion on it, or instruments serving in its movement; yet when he wants to speak he uses a silver instrument, which can in no way serve to this end, [but] rather, hamper utterance. *Item,* he said he was deaf, which is not the case, because we questioned him to find out who had cut off his ear; he answered us by signs that someone had cut it off of him with their teeth.

After said gentlemen of the Bureau had received said report by way of a porter, they had the venerable imposter transported to St. Germain des Prés Hospital and his silver instrument was taken from him. At night he passed over the wall—which is rather high—and from there he took off for Rouen, where he tried to make use of his imposture; which was uncovered, and being apprehended, he was whipped, and banished from the duchy of Normandy on pain of the halter and the Baillif of the poor of this city has assured me of this.

24

ABOUT A HEDGE~WHORE BEGGAR~WOMAN PRETENDING TO BE SICK WITH SAINT FIACRE'S DISEASE, AND A LONG THICK GUT MADE BY TRICKERY CAME OUT OF HER BUM

Monsieur Flecelle [or Flesselles], a Doctor on the faculty of Medicine, a learned man and very experienced, asked me one day to accompany him to the town of Champigny, two leagues from Paris, where he had a cottage. Where, having arrived, while he was walking in his courtyard, a fat wench, very fleshy, came asking him alms in honor of Saint Fiacre, raising her petticoat and her chemise [and] showing a thick bowel, half a foot long and more, which came out of her bum, from which flowed a fluid similar to apostema matter, which had completely stained and besmeared her thighs, together with her chemise, before and behind, so that it was very ugly and foul to look upon. Having questioned her as to how long she had had this disease, she responded to him that it was about four years; then the aforementioned Flecelle, carefully considering her face and the condition of her body, recognized that it was impossible (she being so fat and full-bummed) that such a quantity of excrements could issue forth without her becoming emaciated, dried up, and hectic [wasted away]; and then in one leap he cast himself in great anger on this wench, giving her several kicks below the belly, so much so that he brought her to the ground and made the bowel [or gut] come out of her seat, along with the sound and noise and other stuff, too; and he forced her to declare the imposture to him, which she did, saying that it was an ox's bowel knotted in two places, one of which knots was inside her bum, and said bowel was filled with blood and milk mixed together, in which [bowel] she had made several holes, so that this mixture would ooze out. And immediately recognizing this imposture, he kicked her several more times on the belly, so that she pretended to be dead. Then, he having gone into his house to call one of his servants, pretending to send for some [police] sergeants to take her prisoner, she, seeing the gate to the courtyard open, got suddenly up in a start, just as if she hadn't been beaten, and began running, and never again was she seen in said Champigny.

And still fresh in memory [is the case where] an ugly hedge-whore beggar-woman came along, asking the gentlemen of the Bureau of the poor of Paris to be put on charity, saying that by a bad delivery her womb had fallen, which was the reason why she couldn't earn her living. Then these gentlemen had her examined by the surgeons committed to this charge, and they found that it was an ox's bladder, which was half full of air and smeared with blood, [she] having attached the neck of this bladder deeply into the conduit of her womb, very cleverly, by means of a sponge that she had placed in the extremity of this bladder, which [sponge] being filled, swells and thickens, which was the reason why she could retain it, in such a way that one could not extract it from her, except by force; and thus she could walk without said vessel falling out. Having uncovered this imposture, these gentlemen had her declared prisoner [i.e., arrested], and she didn't get out of prison until the executioner had first chimed carols on her back and afterward she was forever banished from the city of Paris.

25

ABOUT A FAT WENCH FROM NORMANDY, WHO PRETENDED TO HAVE A SNAKE IN HER BELLY

n the year 1561, there came to this city a fat,
full-bummed wench, chubby and shapely, thirty years
old, or thereabouts, who said she was from Normandy,
[and] who was going about to the good houses of ladies and
maidens, asking them for alms, [and] saying that she had a snake
in her belly which had gotten into her while she was asleep in a
hempfield; and she had them put their hand on her belly in order
to get them to feel the snake's movement, which gnawed and
tormented her day and night, as she said; therefore everybody gave
her alms through the great compassion that they had upon seeing
her, in addition to which she put on a good show. Now there was
an honorable lady and greatly charitable who took her into her
dwelling and had me called (together with monsieur Hollier, Regent
Doctor of the faculty of Medicine, and Germain Cheval, Sworn
Surgeon in Paris), to find out whether there could be some way
to drive this "dragon" out of this poor woman's body; and having
seen her, monsieur Hollier ordered a medicine for her that was
rather potent (the which caused her to make several stools) with
the intention of making this creature come out; nevertheless, it did
not come out at all. Drawing together [to consult] again right after
this, we concluded that I should put a speculum into the neck of
her womb; and therefore she was put on a table, where her "badge" 113
was spread, in order to apply the speculum to her, by which I made
a good and ample dilation, in order to find out if one could perceive
head or tail of this creature but nothing was perceived, except a
voluntary movement that said wench made by means of the muscles
of the epigaster: and having recognized her imposture, we drew 114
aside, where it was determined that this movement came from no
animal, but that she was doing it by setting said muscles in action.
And in order to frighten her and know the truth more fully, she
was told that we would renew [our efforts] by giving her another

medicine, much stronger, with the end of making her confess the truth of the matter; and she, fearing to take such a strong medicine again, being sure that she had no snake, took off that very evening, without saying goodbye to her damsel, [but] not forgetting to pack up her things, and a few of the damsel's [also]; and that is how the imposture was uncovered. Six days afterward, I found her at the gate of Montmartre, astride a saddle horse, one leg here and the other there, laughing heartily, and she was going along with the men pushing the fishcarts into town—in order, so I think, to make her dragon fly [i.e., to play her "dragon" or, snake(?) trick on them?] and [then] to return to her country.

Those who counterfeit mutes draw back and double their tongues into their mouths; also those who counterfeit the Saint John's [or, falling] sickness have cuffs put on their hands, and they wallow and plunge in the mire and put blood from some animals on their head, saying that in their thrashing about they have thus injured and bruised themselves; having fallen on the ground, they move their hands and legs about and bestir their whole body, and they put soap in their mouths to make themselves foam, just as epileptics do during their attack. Others make a certain paste with moistened flour and put it over their whole body, shouting that they are sick with Saint Main's disease. Now, a long time ago these beggar-imposters began the practice of fooling the people, for they were already in Asia in the time of Hippocrates, as he writes in his book *On Air and Waters:* wherefore it is necessary to uncover them whenever it is possible, and to accuse them before the Magistrate, to an end that punishment be meted out in proportion as the enormity of the case demands.

26

AN EXAMPLE OF MONSTROUS THINGS DONE BY DEMONS AND SORCERERS

here are sorcerers and enchanters, poisoners and venom-dealers, wicked men, sly men, deceitful men, who carry out their fate through a pact they have made with Demons, [and] who are slaves and vassals to them. And 115
no one can be a sorcerer who has not first renounced God, his creator and savior, and voluntarily made an alliance and friendship with the devil, to whom he has given himself over, in order to recognize and avow him—instead of the living God. And these manners of people who become sorcerers, it is through an infidelity and defiance of God's promises and assistance, or through scorn, or through a curiosity to know secret and future things, or, being oppressed by great poverty, they are aspiring to be rich.

Now no one can deny, and one should not deny, that there be sorcerers; for that is proved by the authority of several Doctors and [other] expounders, old as well as modern, who hold it for a definite thing that there are sorcerers and enchanters, who through subtle, diabolic and unknown means corrupt the body, intelligence, life, and health of men and of other creatures, [such] as animals, trees, grasses, air, earth, and waters. Moreover, experience and reason force us to admit it, because laws have established punishments for such manners of people. Now, one does not make a law for a thing that was never seen or known; for the laws hold cases and crimes that have never been seen nor perceived to be impossible things, and which do not exist at all. Some of them [i.e., witches] were to be found at the nativity of Jesus Christ, and a very long time before this, witness Moses who condemned them at the express command of God, in Exodus, Chap. 22 [and] in Leviticus 19. Ochosias received the death sentence from the Prophet [of God] for having had recourse to sorcerers and enchanters. 116

Devils trouble the intelligence of sorcerers through diverse and strange illusions (see Bodin in his *Republic*), with the result that 117
they think they have seen, heard, said, and done what the devil

pictures to them in their fancy, and [think] they have gone a hundred leagues off, indeed even other things that are completely impossible, not only for men, but also for devils; and this despite the fact that they haven't budged from the bed or [from any] other place. But the devil, since he has power over them, so imprints in their fancy the images of the things that he pictures to them—and that he wants to make them think are real—that they cannot think other than that this is the way it is, and that they have done them and that they have awakened while they were sleeping. Such a thing is done to sorcerers, on account of their infidelity and wickedness [in] that they have given themselves to the devil and have renounced God their creator.

We are taught by Holy Scripture (St. Paul to the Hebrews 1, 14; Gal. 3, 19, 1; Thess. 1, 16) that there are good and bad spirits: the good are called Angels and the bad Demons or Devils. That it be true, [the proof is that] the law is given through the ministry of Angels. Moreover, it is written: Our bodies will be resurrected at the sound of the trumpet and at the voice of the Archangel. Christ says that God will send his angels who will gather the elect from the far corners into Heaven. It can similarly be proved that there are evil spirits, called Devils. That it be thus [the proof is that] in the story of Job (Job, 1, 6), the Devil caused fire to come down from heaven, killed the livestock, raised the winds that shook the four corners of the house, and beleaguered the children of Job. In the story of Ahab there was a spirit of untruth in the mouth of false Prophets (1 Kings, 22). The Devil put into the heart of Judas to betray Jesus Christ. The devils, who were in great number within the body of a single man, were called Legion, and [they] obtained permission from God to enter into swine, which they
118 precipitated into the sea (John 13; Mark 1, 26, 34). There are several other testimonies in Holy Scripture that Angels and Devils exist. From the beginning God created a great multitude of Angels to be citizens of heaven, which are called divine Spirits, and they remain bodiless and are messengers for executing the will of God their creator, either in justice or in mercy; and yet they are diligent [also] in the salvation of men, contrary to evil Angels, called

Demons or Devils, which by their [very] nature always try to hinder the human race with their machinations, false illusions, deceits and lies; and if they were permitted to exercise their cruelty at their will and pleasure, truly, in brief, the human race would be lost and ruined: but they can accomplish only as much as it pleases God to give them leave to do. Which [devils] on account of their great pride were cast out of Paradise and the presence of God; some of which are in the air, others in the water, and which appear above and on the shores, others on earth, others in the deepest [center] of this latter, and will remain until God comes to judge the world; some live in ruined houses and transform themselves into anything they please. Just as one sees many and diverse animals, and other diverse things taking form in the clouds—to wit, Centaurs, snakes, rocks, castles, men and women, birds, fishes, and other things—so do Demons suddenly assume whatever form pleases them; and often one can see them transformed into animals, such as snakes,
toads, owls, dunghill-cocks, crows, he-goats, asses, dogs, cats, 119
wolves, bulls, and others; verily, they take the bodies of humans, living or dead, afflict them, torment them and prevent their natural operations; not only do they transmute themselves into men, but also into Angels of light; they give the semblance of being constrained and of being tied to rings, but such constraint is voluntary and full of treason. These Demons desire and fear, love and scorn; they have charge and office from God to accomplish punishments for the wrongdoings and sins of the wicked, as can be proved [by the fact] that God sent his Writ into Egypt through bad Angels (Numbers 22, 28). They howl at night and make noise as if they were in chains: they move benches, tables, trestles; rock the children, play on the chessboard, turn the pages of books, count money; and one hears them walking about in the chamber; they open doors and windows, cast dishes to the ground, break pots and glasses and make other racket; nevertheless in the morning one sees nothing out of its place, nor is anything broken, nor are the doors or windows open. They have several names, among which *demons,*
cacodemons, incubi, succubi, nightriders, goblins, imps, bad Angels, Satan, 120
Lucifer, father of falsehood, Prince of darkness, legion (Psalms 78; Pierre

de Ronsard in his *Hymnes*); and an infinity of other names, which
121 are written in the book on the *Imposture of Devils,* according to the differences in the evils they carry out and the places where they are most often located.

27

ABOUT THOSE WHO ARE POSSESSED OF DEMONS, WHO SPEAK IN VARIOUS PARTS OF THEIR BODIES

hose who are possessed of Demons speak—their tongues having been torn out of their mouths—through the belly, through the natural parts [genitals] and they speak various unknown languages. They make the earth quake; they make the thunder roll and the lightning flash, and they make the wind blow; they uproot and tear up trees, no matter how big and strong these latter may be; they make a mountain move from one place to another; they raise a castle up in the air and set it back in its place; they charm one's eyes and bedazzle them, with the result that they often make one see that which doesn't at all exist. Which I attest having seen done by a sorcerer, in the presence of the now defunct King Charles IX, and other great lords.

122 Paul Grillant writes of having seen in his times a woman sorcerer, burned in Rome, who would make a dog speak. They do still other things that we will tell hereafter. Satan, in order to teach sorcery to the greatest sorcerers, interlards words from Holy Scripture and
123 from the Doctor-saints in order to make poison out of honey, which has always been and will be Satan's craft. The sorcerers of Pharaoh counterfeited the works of God.

Satan's actions are supernatural and incomprehensible, surpassing the human mind, [it] not being able to explain them, any more than [it can] the magnet which attracts iron and makes the needle urn. And one must not obstinately refuse to accept this truth when ne sees effects and doesn't know their cause, and let us confess he weakness of our mind, without coming to a halt over the rinciples and reasons of natural things—which we do not possess—when we want to examine the actions of demons and enhanters. The evil spirits are the executors and the executioners f God's high justice, and they do nothing except by His permission. Wherefore we must pray God that He not permit that we be led nto the temptations of Satan. God has threatened by His law to xterminate peoples who permitted sorcerers and enchanters to live Leviticus, 2). That is why Saint Augustine in the book of *The City of God* (*Chap*. 20) says that all sects that have ever existed ave meted out punishment to sorcerers, except the Epicureans. Queen Jezebel, because she was a sorceress, Jehu had her thrown out of the windows of his castle and had her fed to the dogs. 124

28

HOW DEMONS INHABIT QUARRIES OR MINES

Louis Lavater writes that those who deal in metals affirm that in certain mines one sees spirits dressed like those who work in mines, running here nd there; and it seems that they are working, whereas they aren't udging; also they say that they do not harm anyone, if no one nakes mock of them, the which happening, they will throw somehing at the mocker, or will hurt him with some other thing. 125

Also, not long ago when I was in the house of the Duke of Ascot, a gentleman of his, named l'Heister—a man of great honor nd who has the chief responsibility in taking care of the duke's ouse—assured me that in certain mines in Germany (in addition o which others have also written this) one used to hear very strange and frightening cries, as if a person were speaking into a po dragging a chain at his feet, coughing and sighing, sometime lamenting like a man who is being racked; another time a nois of a great crackling fire; another time artillery fire, shot from fa off, drums, clarions, and trumpets, the sound of chariots and horse the snapping of whips, the clattering of armor, lances, sword halberds, and other noises as are made in great combats; also noise as when one is bent on building a house, hearing the woo being rough-hewn, the carpenter's line humming, the stone bein cut, the walls being constructed, and other maneuvers, and yet on sees nothing of all that.

The aforementioned Lavater writes that in Davos, in the canto of Grisons, there is a silver mine, which Pierre Briot—a noteworth man and governor of that place—has had worked these last years and he has drawn great spoils from it. There was, among these a spirit which, mainly on Fridays, and often when the miners wer pouring what they had extracted into the tubs, made itself ver busy changing the metals from some of the tubs into others ac cording to its fancy. This governor didn't take heed of it otherwise when he went to go down into his mine, trusting that this spiri couldn't do him any harm, if it were not by the will of God. Now it happened one day that this spirit made a great deal more nois than was customary, so much so that a miner began to say injuriou things to it, along with curses, and to order it to go hang itsel or to go to hell; then this spirit took this miner by the head, whic it twisted on him, in such a way that the front of his head wa directly in the back; and nonetheless he did not die of it, but h lived a long time after, having a twisted neck [and he was] know intimately by several [persons] who are still living; and some year afterward he died.

He writes a lot of other things about spirits, which everyon can read in his book.

Said Louis Lavater, in the above-mentioned book, says that h has heard tell from a prudent and honorable man—the baillif o a lord's manor under the jurisdiction of [the city of] Zurich—wh affirmed that one summer day, early in the morning, going to strol n the meadows, accompanied by his manservant, he saw a man, hom he knew well, wickedly having to do with a mare, which he as very astonished at; he returned quickly and went and rapped n the door of the person he thought he'd seen. Now he found or certain that the other had not budged from his bed; and if this aillif had not diligently found out the truth, a good and honest erson might have been imprisoned and tortured. He relates this story to the end that Judges be well advised [careful] in such cases.

29

HOW DEMONS CAN DECEIVE US

Now these Demons can in many manners and fashions deceive our earthbound heaviness, by reason of the subtlety of their essence and malice of their will; or they obscure the eyes of men with thick clouds that scramble ur minds giddily and deceive us by satanic impostures, corrupting ur imaginations through their buffooneries and impieties. They re [learned] "doctors" in falsehoods—the roots of malice—and n all [kinds of] wickedness [destined] to seduce and deceive us nd prevaricators of truth; and, to say it in a word, they have ncomparable skill in deceits, for they transmute themselves in a housand ways and heap on the bodies of living persons a thousand trange things, such as pieces of cloth, bones, iron instruments, ails, thorns, thread, twisted hair, pieces of wood, snakes, and other nonstrous things, which they often cause to issue through the onduit of women's wombs, which is done after having bedazzled nd altered our imaginations, as we have said.

Some are called *Incubi* and *Succubi:* Incubi are demons which ransform themselves into the guise of men and copulate with omen sorceresses; Succubi are demons that transmute themselves nto the guise of women. And such cohabitation is not done just hile sleeping, but also while awake, which sorcerers and sorceresses have confessed and maintained several times, when they were bein put to death.

Saint Augustine (in *The City of God,* in the 22d and 23d chapter. 15th book) has not at all denied that Devils, transformed into th form of a man or of a woman, can exercise the works of Natur and "do business" [copulate] with men and women, in order t draw them to lust [and] to trick and deceive them, which not onl the people of antiquity have experienced; even in our time this ha happened in several provinces, to various persons with whom th devils, transfigured into men and women, have "done business."

Jacobus Rueff, in his books *De conceptu et generatione homin* (Last Chap., Book 5), bears witness that in his times a lost woma [i.e., a harlot] "did her business" with an evil spirit at night, [i having the face of a man, and that suddenly her belly swelled o her, and thinking she was pregnant, she fell into a strange malad in that all her entrails fell, without her being able to be succore by any physician's or surgeon's skill.

The same thing is written about a butcher's helper who, bein plunged deep in empty musings on lust, was astonished that h suddenly perceived before him a Devil in the form of a beautif woman, with whom, having "done business," his genital parts bega to burn, so that it seemed to him he had a fire burning within hi body, and he died miserably.

Now it is an absurd thing for Pierre de la Pallude, and fo
126 Martin d'Arles, to maintain that devils let the seed of a dead ma flow into the lap of a woman, from whom a child can be engenderec which is manifestly false; and in order to disprove this empt opinion, I shall say only that seed, which is made of blood an spirit [and] which is apt for reproduction, if transported very littl [or slowly], or not at all, is immediately corrupted and altered, an consequently its force is completely extinguished, because th warmth and spirit of the heart and of the whole body is absen from it, so much so that the seed is no longer free of excesses either in quality or in quantity. For this reason, physicians hav judged the man who would have too long a male member to b sterile, because the seed, having had to take such a long journey

already cooled before it is received into the womb. Also, when
he man withdraws too suddenly from his partner, having ejected
is seed, it can be altered in the air which enters into the womb,
hich causes it not to produce fruit. Thus, therefore, one can
ecognize how seriously wrong was Albert the Scholiast, who wrote 127
hat if seed fallen to the ground were put back into the womb, it
ould be possible for it to conceive. One can say the same about
concerning Averroes's neighbor-lady, who (he says) had assured
im by oath that she had conceived a child from a man's seed that
e had ejected into a bath, and having bathed in it, she became
regnant from it. Also, you must not believe at all that Demons
r devils who are of a spirit nature can have carnal knowledge of
omen; for in the execution of that act flesh and blood are required,
hich spirits do not have. Moreover, how would it be possible that
pirits, which have no body at all, might be smitten with love for
omen, and they might engender in them? And also, where there
re no reproductive parts, there is also no coupling; and where
here is no meat or brew there is no seed; also, in cases where it 128
as not been necessary to have offspring and repopulation, Nature
as not granted the desire for engendering. Besides, Demons are
nmortal and eternal; what necessity, then, have they of this re-
roduction, since they have no use for offspring, in as much as
ey [themselves] will always exist. What's more, it is not in the
ower of Satan nor of his Angels to create new ones; and if it
ere so, if Demons had been able to engender other Demons ever
nce they were created, there would surely be a lot of "diablerie"
devils and devilment] at large.

Now as for me, I believe that this cohabitation claimed [by others] is imaginary, proceeding from an illusory impression of Satan.

30

AN EXAMPLE OF SEVERAL DIABOLICAL ILLUSIONS

And in order that one not think that the artific of the Devil is [applicable only to] ancient [times], h has still carried on his practices in our epoch in simila ways, as several [persons] have seen, and many learned men hav written, of a very beautiful girl in Constance, whose name wa Magdaleine, a servant-woman of a very rich citizen of said cit which [girl] published everywhere that the Devil had impregnate her one night, and on this account the Potentates of the city ha her put in prison in order to observe the result of this pregnanc The hour of her delivery having come, she felt the wringing ache and pains habitual to women who are about to deliver; and whe the midwives were about to receive the fruit and thought that th womb should open there began to come out of the body of tha girl iron nails, small pieces of wood and of glass, bones, stones an hair, tow, and several other fantastic and strange things, which th devil through his artifice had put there, to deceive and make foo of the common people, who give credence rather lightly to trick and deceits.

Boistuau affirms that he could produce several other simila stories, related not only by philosophers, but also by ecclesiastic who confess that devils—through the permission of God, or fo the punishment of our sins—can thus abuse men and women, bu [to say] that from such a coupling some human creature can b engendered, this is not only false, but contrary to our religio which holds that there was never any man engendered withou human seed, except the Son of God. Indeed, as Cassianus sai what absurdity, repugnance and confusion there would be in Natur if it were legitimate for devils to conceive by humans, and fo women [to conceive] by them! How many monsters—from th creation of the world to the present—the devils would have brough forth through the whole human race, casting their seed into th wombs of animals, thus creating, through perturbations of seed, an infinity of monsters and marvels!

31

ON THE ART OF MAGIC

oreover, the art of magic is done by the wicked artifice of Devils. Now, there are several sorts of magicians; some make devils come to them, and interogate the dead, which [magicians] are called *necromancers;* others are called] *chiromancers,* because they divine through certain lines hich are in one's hands; others [are called] *hydromancers,* because ey divine through water; others [are called] *geomancers,* because ey divine through earth; others *pyromancers,* who divine through re; others *aëromancers,* or augurers, or prognosticators of how the ature is disposed, because they divine through air, that is, through e flight of birds or through twisters, storms, tempests, and winds. ll of which do nothing but deceive and abuse disbelievers [in hristian faith], who frequent and have recourse to these diviners, rophets, witches, enchanters, [and] who are, more than any other ersons, habitually oppressed by perpetual poverty and dearth, ecause devils engulf them in an abyss of darkness, making them elieve a lie is the truth, through illusions and through false, trouble-rovoking and senseless promises, which is a madness and an nbearable quagmire of error and derision. Above all these men ust be avoided and driven afar by those who know the true eligion, as Moses did by God's commandment.

Jean de Marconville, in his book *Memorable collection of some agical cases that have occurred in our times,* writes of a woman ivineress, a sorceress of Bologna, the big one, in Italy, who, after 129 aving practiced her diabolical art for a long time, fell into a rievous illness, because of which she finished her days. Which eeing, a magician, who had never wanted to disassociate himself rom her] on account of the profit he made from her art during er lifetime, put a certain venomous poison under her armpits, so at by virtue of this poison she seemed to be alive and could be ound in gatherings as she was accustomed [to be], not seeming be different from a living person, except her color which was xtremely pale and wan. Some time afterward, there was another agician in Bologna who took a whim to go see this woman, because she had a great reputation by reason of her art; whic [magician] having arrived at this spectacle, as had others, to watc her ply her craft, suddenly cried out saying, "What are you doin here, gentlemen? This woman, whom you calculate to be doin these fine somersaults and juggling tricks before you, is stinkin and filthy dead carrion!" And all of a sudden she fell to the groun dead, so that the magic of Satan and the abuse of the enchante was manifested to all the persons present.

Langius, in his *Letters on Medicine* [*Medicinalium epistolarum* (Epistle 41), tells of a woman, possessed of an evil spirit, who, afte having been afflicted by a cruel stomach pain, being abandoned b the Physicians, suddenly vomited up some very long and curve nails and brass needles, packed with wax and hairs. And in th same Epistle he writes that in the year 1539, in the village name Tuguestag, a certain plowman named Ulrich Nenzesser, after hav ing endured a cruel pain in the flank, an opening having been mad on him with a razor, a steel nail came out; nonetheless the pair got more and more severe, and, unable to bear it, he slit his throa and having been opened up, they found in his stomach a piece c wood, long and round, four brass knives, some of which were sharp pointed, others toothed in the manner of a saw, and together wit this, two sharp iron instruments which surpassed the length of
130 half cubit and with a large ball of hair. It is seemingly true th
131 all these things were done by the cleverness of the devil, wh deceived the persons present through their [faculty of] sight.

What's more, not long ago I saw an imposter and enchanter– in the presence of King Charles IX, and of Milords and Marecha of Montmorency, and of Retz; and of the lord of Lansac, and c monsieur de Mazille, head physician of the King, and of monsie de Saint Pris, ordinary valet-de-chambre of the King—do sever other things that are impossible for men to do without the cra of the devil, who deceives our vision and makes a false and imagina thing visible to us; which the aforementioned imposter freely co fessed to the King, [namely] that what he was doing came fro the craft of a spirit, and he still had three years to be leashed t him, and that he tormented him greatly; and he promised the Kin that once his time had come and was accomplished, he would be a good man. May God be willing to grant him the grace, for it is written, "Thou shalt not suffer a witch to live." King Saul was cruelly punished for having sought counsel of a woman-enchantress [Endor]. Moses likewise commanded his Hebrews to put forth every effort to exterminate the enchanters around them. (Exodus, 20th Chap., Leviticus 19, I Kings, 28, Deuteronomy.) 132

(In the city of Karenty, the men having called the women to lie 133
with them had the custom of locking themselves to them in the manner of dogs, and they couldn't detach themselves from them for a long time; and having sometimes been found, they were condemned by law to be hanged on a perch back side to, and tied by an unusual line, and they served the people as a spectacle of ridicule; and such a thing, which was a detestable mockery, was done
through the devil's handiwork. [1573, 1575].) 134

32

ON CERTAIN STRANGE ILLNESSES

Now in order to content the mind of the reader 135
still more concerning the imposture of devils and of their slaves—magicians, witches, charmers, and sorcerers—I have collected these stories from Fernel (ex. Chap. 16, Book 2 of *De abditis rerum causis,* Fernel), just as follows.

There are some illnesses that are sent to men by God's permission, and these cannot be cured by ordinary remedies, which for this reason are said to surpass the ordinary course of illnesses with which men are accustomed to being afflicted. Which can be easily proved by Holy Scripture itself, which gives us to believe that for the sin of David there occurred such a corruption of air that the plague cut off the thread of life of more than sixty thousand persons. We read also in the same Scripture that Ezechias was afflicted with a very great and very grievous illness. Job received so many ulcers on his body that he was totally covered with them;

which happened to them through the permission of this great God who governs this lower world, and all that is contained within it, according to His will.

Now just exactly as the Devil, chief and sworn enemy of man, often (yet through God's permission) afflicts us with great and diverse maladies, so do sorcerers, tricksters, and wicked men—through ruses and diabolical tricks—torture and abuse countless men; some invoke and adjure heaven knows what spirits, through whispers, exorcisms, imprecations, enchantments, and bewitchments; others tie around the neck—or else carry on them in some other way—certain writings, certain characters, certain rings, certain pictures, and other such claptrap; others use certain harmonious chants and dances. Sometimes they use certain potions, or, rather poisons, suffumigations, perfumes, charms, and enchantments. Some are found who, having contrived the image and likeness of some absent party, pierce it with certain instruments, and boast of afflicting—with any such illness as pleases them—the one whose likeness they are piercing, even though he may be far away from them; and they say that this is done by virtue of the stars and of certain words that they hum while piercing such an image or likeness made of wax. There are, in addition, an infinity of such villainies which have been invented by these rascals to afflict and torment men, but it would weary me to say any more about it.

There are some of them who use such witchcraft to keep men and women from consummating their marriage, which in vulgar
136 language one calls "knotting the point of the codpiece." There are some of them who prevent a man from urinating, which they call *cheviller,* or "pegging or pinning him up." There are some of them also who by their sorceries make men so unable to sacrifice to Madame Venus that the poor women who are trying to "do business" with them think they are castrated, and more than castrated.

Such riffraff not only afflict men with several and diverse sorts of maladies; but also, hangmen and sorcerers that they are, they hurl devils into the bodies of men and women. Those who are thus tormented with devils by the sorceries of these rascals differ in no way from simple maniacs, except that they say things that are marvelously great. They tell everything that has occurred before, even though it might be very well hidden and unknown, except to very few people. They discover the secret of those who are present, insulting them and reproving them so energetically that they would be more than dullards if they didn't feel it; but just as soon as anyone speaks of the Holy Scripture, they are completely terrified, they tremble and are very vexed.

Not long ago an individual, on account of the great summer heat, got up during the night to drink, who, not finding any fluid to quench his thirst, takes an apple that he spies, [and] who, as soon as he had bitten into it, it seemed to him that he was being suffocated; and already, like one beseiged by an evil spirit hidden in this apple, it seemed to him that, in the middle of the darkness, he saw a big, very black dog who was devouring him, who, having thereafter gotten well, told us from start to finish everything that had happened to him. Several Physicians having taken his pulse, having recognized the extraordinary heat there was in him, [along] with a dryness and blackness, from which they opined that he had a fever [ague] and inasmuch as he didn't rest at all, and since he didn't cease to dream, they judged him to be out of his mind.

A few years ago a young Gentleman over an interval of time fell into a certain convulsion, sometimes having only the left arm, sometimes the right, sometimes a single finger, sometimes the spine, the back, and the whole body so suddenly jerked and tormented by this convulsion that four valets could hold him in his bed only with great difficulty. Now, the fact is that his brain was not at all agitated or tormented: his speech was unimpeded, his mind not at all troubled, and all his senses [were] intact, even at the height of the convulsion. He was wracked two times a day at least by such a convulsion, having come out of which he felt well, except that he was very tired and spent because of the torment that he had suffered. Any well-informed Physician would have been able to judge that it was a true epilepsy if, along with this, the senses and the spirit had been troubled. All the finest Physicians having been summoned there, they gave the opinion that it was a convulsion very closely approximating epilepsy, which was aroused by a malignant vapor, enclosed within the spine of the back, without its in any way harming the brain. Such opinion having been settled upon as the cause of this illness, nothing was omitted from what is known to the art in an effort to bring relief to this poor patient. But in vain did we put forth all our efforts, being more than a hundred leagues away from the cause of the illness. For, three months later it was discovered that it was a devil who was the author of this sickness, and he himself declared himself, speaking Greek and Latin copiously through the mouth of the patient, although said patient knew nothing about Greek. He uncovered the secret of those who were present, and especially of the Physicians who were present, making mock of them because with great peril he had entrapped them and because with their useless medicines they had almost caused the patient to die. Each and every time his father went to see him, as soon as he spied him from a distance, he shouted "Make him go away; keep him from coming in, or else take off the chain he has at his neck," for like the Knight that he was, according to the custom of French Knights, he wore the collar of the order, at the end of which was a likeness of Saint Michael. Whenever one read something from Holy Scripture before him, he bristled, got up, and writhed a great deal more than before. When the paroxysm had passed, he remembered everything he had said or done, repenting of it, and saying that he had either said or done it against his will. This Demon, constrained by ceremonies and exorcisms, said that he was a spirit and that he was not damned for any crime. Being questioned as to what he was or by what means and through the power of whom he was thus tormenting this gentleman, he answered that there were a lot of domiciles in which he hid and that during the time when he let the patient rest, he went off to torment others. [He said] besides that he had been cast into the body of this gentleman by an individual whom he didn't wish to name and that he had entered into that body through the feet, creeping clear to the brain, and that he would go out through the feet when the day agreed upon between the two of them had come. He discoursed on a lot of things according to the custom of the possessed; [and I am] assuring you that I'm not showing forth something new here, but in order that people know that sometimes devils enter our bodies and that they vex them with torments never before heard of.

Sometimes also they don't enter into [the body] but [rather]
agitate the good humors of the body, or else they send bad ones
[humors] to the principal parts, or else they fill the veins with these
bad humors, or plug up the pipes of the body with them, or else
they change the construction of the instruments [organs], as a
result of which countless illnesses occur. Devils are the cause of
all these things; but sorcerers and wicked men are the serfs and
ministers of devils. Pliny writes that in his time Nero lit on the
falsest kinds of magic and sorceries which have ever been. But 137
what's the need of writing about Ethnics [i.e., heathens], given that
Scripture itself bears witness, as is manifest from what is written
about the Pythoness, about the woman-ventriloquist, about King
Nebuchadnezzar, about the sorcerers and enchanters of Pharaoh,
and even about Simon the Magician in the times of the Apostles? 138
The same Pliny writes that a man named Demarchus was changed
into a wolf, having eaten the entrails of a sacrificed child. Homer 139
writes that Circe changed Ulysses's companions into swine. Several
ancient poets write that such sorcerers made fruit [crops] pass
from field to field and from garden to garden. Which does not
seem to be a fable, in as much as the law of the twelve tablets
establishes and ordains certain punishments for such charlatans and
scoundrels.

Now, just as the devil cannot proffer real things, which he could in no way create, so does he proffer only empty [fake] species of these [things], through which he clouds men's minds; thus in illnesses he cannot give a true and complete healing but rather he uses only a false and palliative cure.

I have also seen jaundice disappear from [all over] the surface of the body in a single night, by means of a certain little brief [i.e., short message] that was hung around the neck of the icteric. I have likewise seen agues be healed through word formulas and certain ceremonies, but they returned afterwards, much worse.

There are many other [such remedies] of quite another strain;
for there are ways of doing which we call superstitions, in as much
as they are not founded on any reason or authority, either divine
or human; rather on some old wives' ravings. I ask you, is it not
a true superstition to say that he who carries the name of the three
kings who came to adore our God, to wit, Gaspar, Melchior, and
Balthazar is [or, can be] cured of epilepsy? Which, nonetheless,
the most approved remedies ordinarily do not do, such as, say,
140 essence of *succinum* or amber, mixed with peony conserves, a dose
the size of a hazelnut being given to the patient every morning.
That teeth are mended, if, while mass is being said, one proffers
141 these words: *Os non comminuetis ex eo?* That one can appease
vomiting by certain ceremonies, just by knowing the name of the
patient?

I saw someone stop blood from any part of the body whatsoever,
[just] by humming some words or other. There are some [persons]
142 who say these words: *De latere eius exiuit sanguis et aqua.*

How many such ways are there for curing agues? Some [per-
sons], holding the hand of the fever-ridden patient, say: *Aequè*
143 *facilis tibi febris haec sit, atque Mariae virgini Christi partus.* Others
144 secretly say this beautiful psalm: *Exaltabo te Deus meus rex.* If
someone (says Pliny) has been bitten by a scorpion, and if in
passing he should say it into an ass's ear, he is immediately cured.
Those are some fines ways to cure! Now just as they are cured
through such words, so also through similar and superstitious writings are they cured. As, in order to cure eye disease, there are
some [persons] who write these two Greek letters, π and α, and
wrap them up in a linen cloth and then hang them around their
neck. For toothache they write: *Strigiles falcesque dentatae dentium*
145 *dolorem persanate.*

One can also find great superstition in external applications. Like
the one of Apollonius: namely, to scarify one's gums with the tooth
146 of a man who has been killed in order to be cured of toothache;
[or such] as making pills from the skull of a hanged man, [as an
antidote] for the bite of a rabid dog. As, they [also] say that
epilepsy is cured by eating some flesh of a wild beast which has

been killed with the same weapon that a man has been killed with.
As they also say that a quartan ague is cured if one drinks some 147
wine in which has been dipped a sword with which a man's throat
has been slit. If that were true, the executioner of Paris would be
better off than he is. They say also that to be cured of a quartan
ague one need only place the shavings from one's nails in a linen
cloth, tie them around the neck of a live eel, and throw it immediately back into the water. To cure diseases of the spleen (they
say), one need only place on top of it the spleen of an animal and
have the physician say that he is effecting a cure for the spleen. 148
To cure a cough, one need only spit into the mouth of a red frog,
and let it go free right away. The rope with which someone has
been hanged, tied around the temples, cures headache. It is amusing
to hear such a way of practicing medicine; but among others, this
one is nice: which is to apply this elegant word *Abracadabra* with
a certain flourish, which Serenus writes [of], in order to cure the
ague. It's another fine touch to say that the leaf of Cataputia, [if] 149
plucked upwards causes vomiting, and [if] broken downward causes
the bowels to discharge. And what's more, they have been so
impudent as to claim that there were some herbs dedicated and
consecrated to devils, as Galen relates concerning a certain Andrew,
as also Pamphile (Galen, in the sixth book on *Simples*).

I would never have been finished if I'd [just] wanted to amuse
myself by stringing together thousands upon thousands of [examples 150
of] such superstitious gibberish and I would not have gone on about
it so long, except in order to give warning to a lot [of persons],
who are mistaken about it, not to believe in it any longer, and to
beg them to reject all such foolishness, and to stop at what is
assured, and [this] by so many skillful and worthy gentlemen [who
are] confirmed and certified in Medicine; which doing, an infinite
good will be brought about for the public; all the more because
next to the honor of God, there is nothing that should be more
precious to man than his health. And one should in no wise trust
men who have abandoned the natural means and virtues bestowed
that God has put in plants, animals, and minerals for the curing
of illnesses, and [who] have cast themselves into the nets of evil

spirits who are waiting for them at the pass; for one must not at all doubt that, since they do not rely on the means that God has ordained, and since they abandon this rule, universally established at the creation of the world, one must not be ignorant of the fact that the evil spirits may have taken great pains to hold them to it, handing them one ripe [berry] along with two green ones; and [one must not] rely in this way upon the virtue of words and characters and other kinds of blandishments and gulling, like sorcerers; and they have gone so far as to say that they don't care who may cure them, even if it were the devil from hell, which is a proverb [saying] unworthy of a Christian, for Holy Scripture expressly forbids it. It is certain that sorcerers cannot cure natural illnesses nor can Physicians [cure] illnesses coming from spells.
151 And as for a few Empiricals who cure simple wounds merely by the application of dry linen cloths, or cloths dipped in pure water, [if] sometimes they cure them, one must not believe on account of that that it is an enchantment or miracle, as idiots among the populace think; but by the beneficence of Nature alone, who cures wounds, ulcers, fractures, and other illnesses; for the surgeon does nothing but to help her in some things, and to remove what might hinder her, such as pain, fluxion, inflammation, apostema, gangrene, and other things which she is unable to do, [such] as to reset fractures and dislocated bones, plug up a large vessel in order to stop a flow of blood, extirpate a growth, extract a large stone from the bladder, remove a superfluous piece of flesh, abate a cataract, and countless other things that Nature cannot do on her own.

33

ON INCUBI AND SUCCUBI ACCORDING TO PHYSICIANS

hysicians hold that *Incubus* is a sickness in which the person thinks he is being oppressed and suffocated by some heavy load on his body, and it comes principally at night; the common people say that it is an old woman who is loading down and compressing the body [and] the common people
call her *chauche-poulet.* 152

The cause is most often from having drunk and eaten much too vaporous viands, which have caused an indigestion, from which [viands] great vapors have arisen in the brain which fill one's ventricles, by reason of which the animal faculty—which makes [us] feel and move—is prevented from coming into full luster by the nerves, from which an imaginary suffocation arises, through the lesion which is created as much in the diaphragm as in the lungs and other parts which are used in respiration. And then the voice is impeded, and they [i.e., those suffering from this ailment] have so very little [voice] left [that] it is [nothing but] groaning and stammering, and—if they can speak at all—[only] requesting aid and succor. For curing this, one must avoid vaporous foods and strong wines, and in general all things which are the cause of making fumes rise to the brain.

34

ON "POINT~KNOTTERS"

notting the point; and the words have no effect,
but it is the devil's craft; and those who knot it cannot 153
do it without having convened with the devil, which is a damnable wickedness. For the one who practices this cannot deny that he is a violator of the law of God and of nature, to [thus] prevent the law of marriage, ordained by God. By this [spell] it happens that they cause marriages to be broken, or at the least

they keep them in sterility, which is a sacrilege. Moreover, they remove the mutual friendship from marriage, as well as human society, and they put a capital hate between the two spouses; likewise they are the cause of the adultery and lechery which result from it, for those who are tied [in matrimony] burn with cupidity for one another. Moreover, from this there often come several murders, committed on the person of those that one suspects of having "knotted the point," who very often hadn't [even] thought of it. Also, as we have said above, sorcerers and poisoners, by subtle, diabolical and unknown means, corrupt the body, the life, the health, and the good judgment of men. Wherefore, there is no punishment so cruel which might suffice to punish sorcerers; all the more because all their wickedness and all their designs fly in the face of God's majesty, to scorn him and to offend the human race in a thousand ways.

OTHER STORIES NOT OFF THE SUBJECT

154 ome [persons] estimate that it is a monstrous thing to wash one's hands with melted lead: even Boistuau in his *Histoires prodigieuses,* chapter eight, relates that Jerome Cardan, book six, *De subtilitate* [*rerum*], writes this story—as being miraculous—about it.

"When," says he, "I was writing my *Book of subtle inventions,* I saw a fellow in Milan who washed his hands with melted lead, and he was taking a crown in money from each spectator." Cardan, trying to seek out the secret of this in Nature, says that by necessity it had to be that the water in which he washed his hands first must have been extremely cold and that it must have had some obscure and thick property; nonetheless he doesn't describe it at all.

Now not long ago I found out what it was from a gentleman who held it to be a great secret, and he washed his hands in melted lead in my presence and in that of several other [people], which I marvelled at; and I heartily begged him to tell me the secret, which he gladly granted me, on account of a favor I had done him; the aforementioned water was nothing else than his own urine, with which he washed his hands first, which I have found to be the truth because I have tested it out since. Said gentleman in place of his urine, rubbed his hands with *unguentum aureum,* or with some other similar [unguent], which I have likewise tested, and one can explain it, by the fact that their thick substance prevents the lead from sticking to the hands, and drives it to one side and the other in small clots [or, beads]. And on account of the affection he had for me, he went even further: he took a shovel of red hot iron, and he threw it on top of some slices of bacon and made it melt, with the drippings all aflame he washed his hands; which he told me he did by means of onion juice with which, ahead of time, he had washed his hands.

I wanted very much to relate these two stories (although they are not exactly on the subject) so that some good companion might in this way be able to make an impression among those who wouldn't know this secret.

35
CONCERNING MARINE MONSTERS

t must not be doubted that just as one sees several monstrous animals of diverse shapes on the earth, so also are there many strange sorts of them in the sea, some of which are men from the waist up, called Tritons, others [are] women, called Sirens, [or, Mermaids], who are [both] covered with scales, as Pliny describes them (Pliny, 9th Book of
his *Natural History*), without, nonetheless, the reasons which we 155
have brought to bear before, regarding the fusing and mixing of seed, being able to apply in the birth of such monsters. Besides, one sees in rocks and plants, effigies of men and other animals, and there is no explanation for them, except to say that Nature is disporting herself in her creations.

43. Picture of a Triton and a Siren, seen on the Nile

In the time when Mena was governor of Egypt, taking a walk one morning on the shore of the Nile, he saw a man coming out of the water up to the waist, his face stern, his hair yellow, mixed through with a few gray hairs, his stomach, back, and arms well-formed, and the rest of a fish. The third day after, toward break of day, another monster also appeared out of the water with the face of a woman; for the softness of the face, the long hair, and the breasts showed that well enough; and they remained so long above the water that everyone in the city saw both of them to their full content.

Rondelet, in his book *On Fish,* writes that a marine monster was seen on the Norwegian sea, which, as soon as it was caught, everyone gave it the name of Monk and it was such.

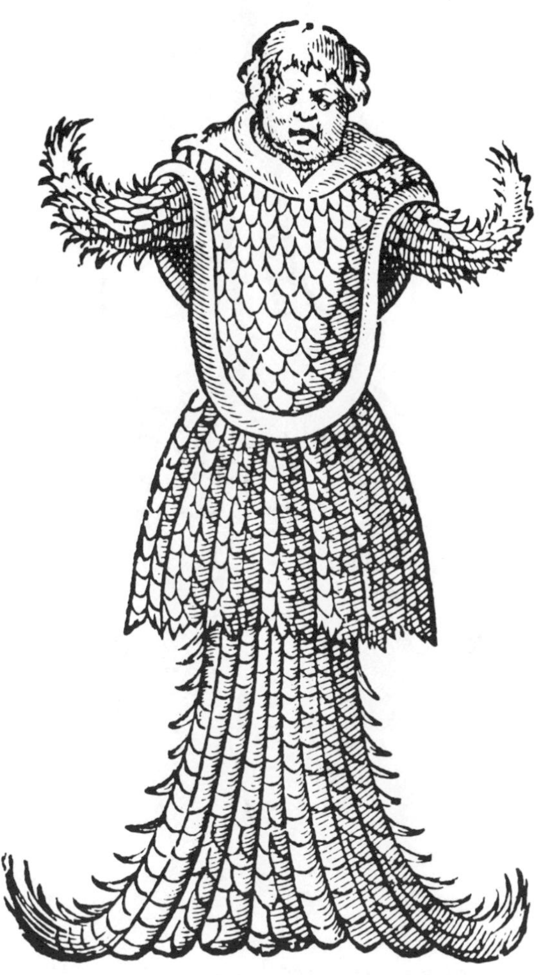

44. Marine monster having the head of a Monk, armed and covered with fish scales

Another monster described by said Rondelet, in the manner of a Bishop, covered with scales, having his miter and pontifical ornaments, which was seen in Poland, in 1531, as Gesnerus describes. [Fig. 45]

Hieronymus Cardanus sent this monster to Gesnerus, which had a head similar to a bear, arms and hands almost like a monkey, and the rest of a fish, and it was found in "Macerie." [Fig. 46] 156

45. Figure of a marine monster resembling a Bishop dressed in his pontifical garments

In the Tyrrhenian Sea, near the city of Castre, this monster was caught, having the form of a lion covered with scales, which was presented to Marcel, then Bishop, who, after the death of Pope Paul III succeeded to the Papacy. This Lion had a voice similar to that of a man; and with great admiration it was led into the city, and soon afterward it died, having lost its natural habitat; as Philippe
157 Forestus testifies to us in Book 3 of his *Chronicles*.

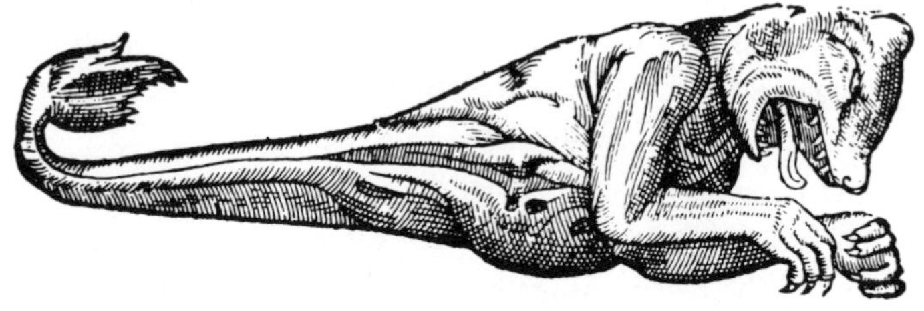

46. Figure of a marine monster having the head of a Bear and the arms of a Monkey

47. Figure of a Marine Lion covered with scales

In the year 1523, on the third day of November, this marine monster was seen in Rome, the size of a child of five or six years of age, having a human upper half as far as the navel—except the ears—and the lower [half] similar to a fish.

Gesnerus mentions this marine monster whose picture he had
obtained from a painter who had seen it in its natural setting in
Antwerp, having a very savage head, with two horns, and long ears,
and all the rest of the body that of a fish, except the arms which
158 approached the normal; which was caught in the Illyrian Sea,
throwing himself up on to the shore, trying to capture a small child
who was near there, and, being hotly pursued by some sailors who
had caught sight of it, it was wounded with blows from stones and
shortly afterward came to the water's edge to die.

This marine monster having the head, mane and forequarters
159 of a Horse, was seen in the Ocean sea; the picture of which was
brought to Rome, to the Pope then reigning.

160 Olaus Magnus says he has seen this marine monster of an
English Gentleman; and it had been caught near the shore of

48. Picture of a Marine monster having a human torso

49. Hideous figure of a Sea Devil

50. Figure of a Sea Horse

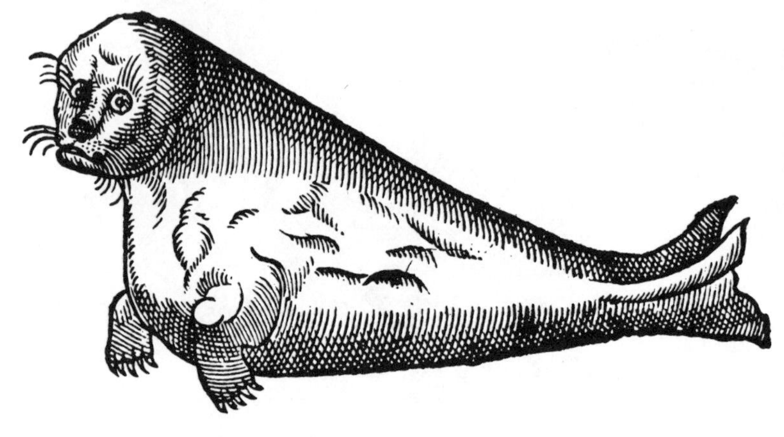

51. Figure of a Marine Calf

161 Bergen, which [monster] ordinarily lived there. Again not long ago
a present was made of a similar one to the now defunct King
(Charles IX, King of France), which he had [his people] nourish
for quite a while in Fontainebleau, which often came out of the
162 water and then went back in it.
This marine monster, as Olaus [Magnus] says, was seen in the
163 sea, near the Isle of Thylen, located toward the North, in the year
of grace 1538, of an almost unbelievable size, to wit, seventy-two
feet long, fourteen feet high, having a distance of seven feet or
thereabouts between its two eyes; its liver was so large that it filled
five wine casks; its head [was] similar to a Sow, having a crescent
located on its back, three eyes in the middle of each side of the
body, and the rest completely covered with scales, as you can see
164 by this figure.
165 The Arabs inhabiting Mount Mazovan, which is along the Red
Sea, ordinarily live on a fish named Orobon from nine to ten feet
long, and wide in accordance with and in proportion to its length,
having scales made like those of the Crocodile. This one is won

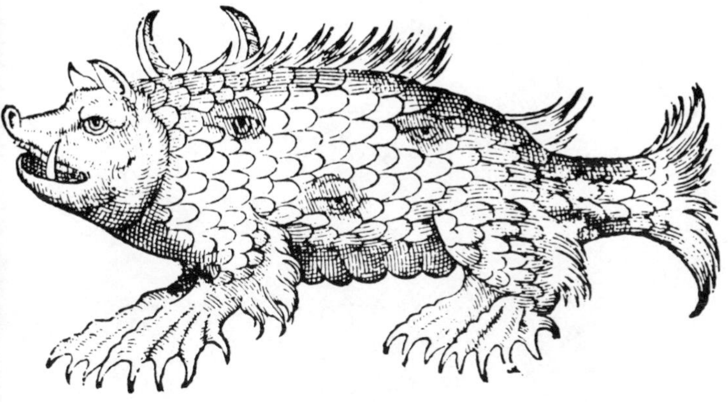

52. Figure of a Marine Sow

drously ferocious against other fish. André Thevet makes a rather full declaration concerning it in his *Cosmography,* from which I took this portrait, as of a very monstrous animal. [Fig. 53]

The Crocodile as Aristotle writes in the books on the history and qualities of animals, is a large animal, fifteen cubits long. It does not bring forth a [live] animal, but eggs, no fatter than those of a goose; it lays at most sixty of them. It lives a long time, and from such a small beginning comes such a large animal; for the little hatchlings are in proportion to the egg. It has such an impedite [rudimentary] tongue that it seems not to have one at all, which is the reason why it lives partly on land, partly in the water; as, being terrestrial, it takes the place of a tongue for him, and as, being aquatic, he is without a tongue. For fish, either they have no tongue at all, or they have one that is very tied and impedite. Only the Crocodile among all animals moves the upper jaw; the lower jaw remains stable, because its feet cannot serve it to catch or to hold. (The parrot moves its upper and lower beak.) It [the crocodile] has eyes like a hog, long teeth that come out of its snout, very pointed claws [ungues] and hide so tough that there 166

53. Figure of a fish named Orobon

is no arrow or pointed weapon that could pierce it. They make a medicine from the Crocodile called *Crocodilea,* for suffusions [or dimness of sight] and cataracts of the eyes; it cures specks, splotches, and pimples that come on the face. Its gall, when applied to the eyes, is good for cataracts; its blood, applied to the eyes,
167 clarifies one's sight.

Thevet, in his *Cosmography,* vol. 1, chap. 8, says that they live in the fountains of the Nile, or in a lake which comes from said fountains, and he says that he saw one of them that was six paces long, and more than a good three feet wide over the back, so much so that its appearance alone is hideous. The manner of catching

54. Catching Crocodiles

them is as follows: As soon as the Egyptians and Arabs see that the water of the Nile is receding, they toss out a long rope, at the end of which there is an iron hook, rather thick and wide, weighing about three pounds, to which they attach a piece of camel flesh, or [flesh] of some other animal; and when the Crocodile sees the prey, he doesn't fail to throw himself upon [it] and swallow it down; and the hook having been devoured, [and he] feeling pricked [or, stuck], it is delightful to see him leaping into the air, and into the water. And when he is caught, these barbarians pull him little by little up near the edge of the shore, having placed the cord over a palm tree or some other tree, and thus they suspend him somewhat in the air, for fear that he will throw himself upon them and devour them. They give him several blows with a wooden bar and they beat him to death, and kill him; then they skin him and eat his flesh, which they find very good.

Jean de Lery, in chapter 10 of his *History of Brazil,* says that the savages [natives; Indians] eat Crocodiles, and that he has seen live offspring [of crocodiles] brought by the savages to their houses, around which their little children play, without their hurting them at all.

168 Rondelet, in his book on Insect fish, that is to say [fish] which are by nature half way between plants and animals, offers these two figures, one called Sea Panache [plume], because it resembles the plumes that people wear on their hats: fishermen—on account of the similarity it has to the end of the male member—call it *Vit-*
169 *volant* [flying-prick]; when alive, it swells up and makes itself thicker;

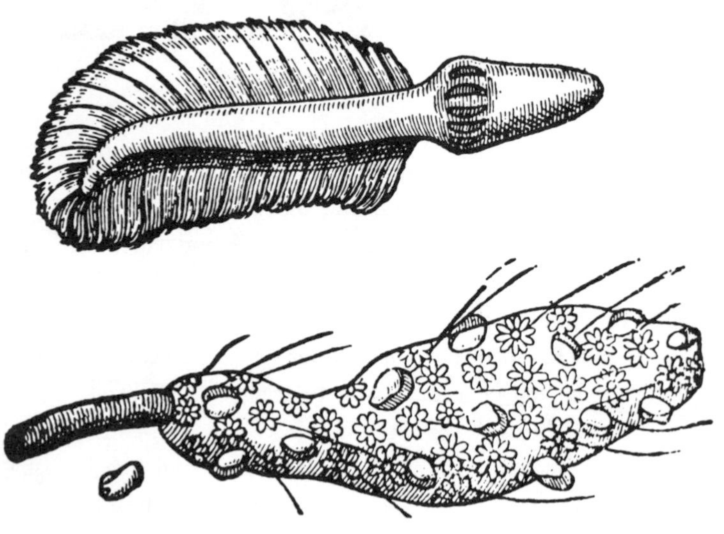

55. Two fish, one like a plume and the other like a bunch of grapes

being deprived of life, it becomes totally withered and limp. It shines at night like a star.

Pliny writes that in the sea one finds not only forms of animals that are on the land; but I believe that this picture is the cluster [of grapes] of which he is speaking; for all over the bottom it resembles a bunch of grapes which is in flower; it is long, like a formless mass, hanging from a tail. 170

In the sea of the Spanish Isle [Hispaniola?], in the new lands, several [kinds of] monstrous fish are found. Among which Thevet, book 22, chap. 12, Vol. 2 of his *Cosmography,* says he has seen a very rare one that in the language of the country they call Aloés, and it is similar to a goose, having its neck raised high, its head to a point like a Good-Christian pear, the body thick like that of 171

56. The "Aloés," a monstrous fish

a goose, without scales, having its four fins under the belly, and if you saw it above the water you would say it was a goose [behaving like any waterfowl], bobbing among the waves of the sea.

172 The Sarmatian Sea, which is otherwise called the Eastern Germanic Sea, nourishes so many fish unknown to those who live in warm regions, and so monstrous, that [there is] nothing more so.
173 Among others, there is one made exactly like a snail, but thick as a wine cask, having horns almost like those of a stag, at the end of which and on the branches of which there are little balls [or bulbs], round and lustrous, like fine pearls. It has a very thick neck; its eyes shine like a candle, its nose is roundish and made like that of a cat, with a little bit of hair all around [it], having a very wide-slit mouth, beneath which a projection of flesh hangs on it [that is] rather hideous to see. It has four legs, and wide, hooked paws which serve it as fins, with a rather long tail, all spotted and colored

57. Snail from the Sarmatian Sea

in various colors, like that of a tiger. It stays out at high sea because it is timorous: for I am assured that it is amphibious, partaking of water and of land. When the weather is clear, it plants itself on the sea shore, where it grazes and eats what it finds [to be] best. Its flesh is very delicate and pleasing to eat; whose blood is good for those who have diseases of the liver or lungs, as is that of the giant turtles for those who are afflicted with leprosy. Thevet says he got this from the country of Denmark. (Thevet, book 20, chap. 18, vol. 2 of his *Cosmogr.*)

In the huge, deep, fresh water lake—on which the large city
of Themistitan, in the Kingdom of Mexico, is built on pilings, like 174
Venice—is found a fish as big as a sea-calf. The savages of the Antarctic call it Andura; the barbarians of the country and the Spaniards—who have made themselves masters of this place by conquests of their new lands—call it Hoga. Its head and ears are not very different from a terrestrial swine; it has five whiskers a half-a-foot long or thereabouts, similar to those of a big Barbell; its flesh is very good and delicious. This fish produces live offspring, in the fashion of a whale. If you contemplate it while it is disporting itself swimming in the water, you would say that it is now green, now yellow, and then red, just like the Chameleon; it keeps more to the edge of the lake than elsewhere, where it feeds on leaves of a tree called Hoga, from which it took its name. It is very toothy and savage, killing and devouring other fish, indeed [those] bigger than it is; that is why people pursue it, hunt it and kill it, because
if it entered into the conduits it wouldn't leave a single one of them 175
alive; whereby the person who kills the most of them is most welcome. Which is written by Thevet, chapter 22, volume 2, of his *Cosmography.* [Fig. 58]

André Thevet, vol. 2, of his *Cosmography,* chapter 10, while swimming in the sea, says that he saw an infinite store of flying fish that the savages call Bulampech, who leap so far out of the water from which they are coming that one can see them land fifty feet from there, which they do in as much as they are pursued by other large fish which seize their catch from them. This [kind of] fish is small like a mackerel (I have one of them in my office that

58. Portrait of the Hoga, a monstrous fish

was given to me and that I keep as a memento), having a roun head, a back of a bluish color, and two wings, almost as long a the whole body, which it hides under its jaws, [the wings] bein made just like the baleens or ailerons with which other fish hel themselves swim. They fly in rather great abundance, chiefly a night, and while flying, they bump against the sails of ships an fall within. The savages take nourishment from their flesh.

Jean de Lery in his *History of Brazil,* chapter 3, confirms this
and says that he saw big droves of fish come out of the sea an
rise into the air (just as on earth one sees larks or starlings) flyin
almost as high out of the water as a pike, and sometimes as fa
176 as two hundred and fifty to five hundred feet. But also it has ofte
happened that some of them, bumping against the masts of ou

59. Portrait of certain flying fishes

hips, [and] falling within, we would catch them by hand. This fish
s in the form of a herring, yet a little longer and thicker; it has
ittle barbels under the throat, and wings like a bat and almost as
ong as the whole body; and it tastes very good and is very delicious
o eat. There is yet another thing (says he) that I have observed: 177
This is that neither within the water nor out of the water are these
oor flying fish ever at rest; for being in the sea, the big fish pursue
hem in order to eat them, and wage perpetual war on them; and
f in order to avoid this they try to save themselves in the air and
n flight, there are certain marine birds who catch them and feed
n them.

Between Venice and Ravenna, one league [two miles] above
Quioze in the sea of the Venetians in the year 1550, a flying fish 178

was caught that was frightening and gave marvel to see, [being four feet and more in length and twice as much in width from on tip to the other of its wings, and a good square foot in thickness Its head was wondrously thick, having two eyes, one on top of th other, in a line; two large ears and two mouths; its snout was ver fleshy, green in color; its wings were double; on its throat it ha
179 five holes in the fashion of a Lamprey; its tail was an ell long, o top of which were two little wings. It was brought quite alive t

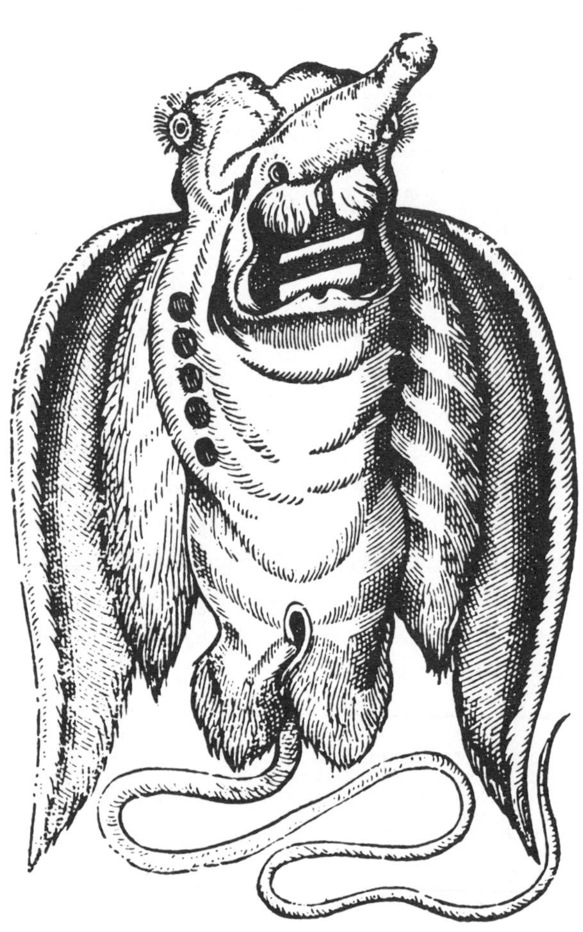

60. Figure of another very monstrous flying fish

aid city of Quioze, and presented to the lords of this latter, as a hing that had never [before] been seen.

There are found in the sea such strange and diverse sorts of hells that one can say that Nature, chambermaid of the great God, isports herself in the manufacture of them, among which I have ad these three portrayed for you, which are worthy of great ontemplation and wonderment, in which there are fish [that are] ke snails in their shells, which Aristotle, book 4 of the *History f Animals,* names Cancellus, being companions of crusted and ard-shelled fish, and similar to spiny lobsters, being born without hells. [Fig. 61]

Rondelet, in his book of the *History of Fish,* says that in Languedoc his fish is called *Bernard the Hermit:* it has two longish and slender orns under which are its eyes, not being able to draw them in as o crab fish [or, crayfish], but they always appear protruding: its orefeet are cloven and forked, which serve it in defending itself nd in carrying [things] in its mouth. It has two others, curved and ointed, with which it helps itself move. The female makes eggs, hich one can see hanging on her rear, like little threaded beads; nd yet [they are] enveloped and bound by little membranes.

Elian in book 7, chapter 31, writes the following concerning it:

> Cancellus is born completely naked and without a shell, but after some time, it chooses an appropriate one to dwell in, when it finds empty ones, such as those of the Purple Shellfish, or of some other one found empty; it takes up residence there, and having become bigger, so that it cannot stay there any longer, (or when Nature incites it to spawn), it seeks out a bigger one where it lives with plenty of room and in comfort. Often there is combat among them over entering [into a given shell], and the strongest casts out the weakest and enjoys the place.

Pliny, book 9, gives the same testimony. 180

There is another fish named Pinothera (Plutarch), of the crab-ish sort, which maintains itself and always lives with the Pinna

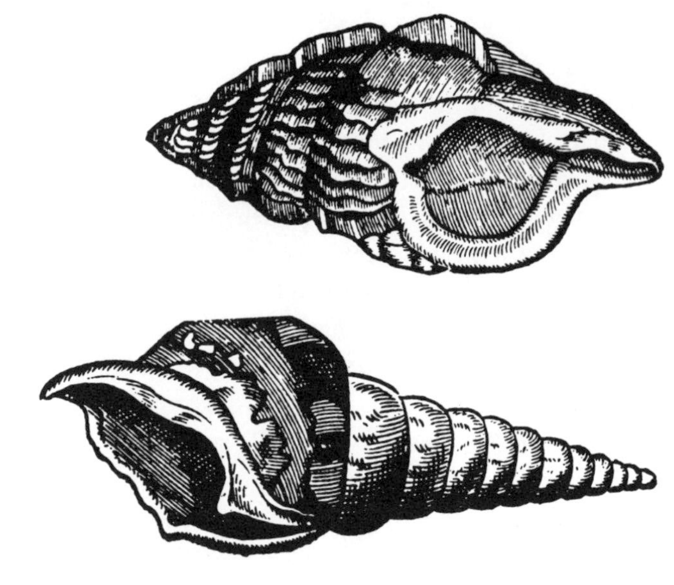

61a.

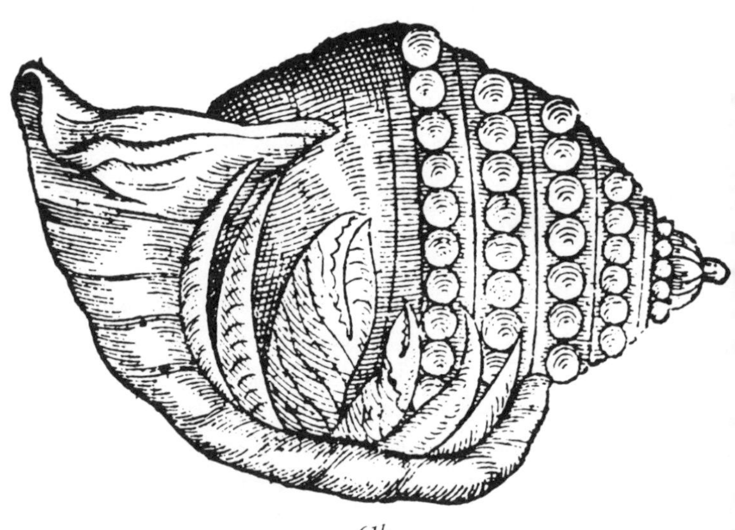

61b.

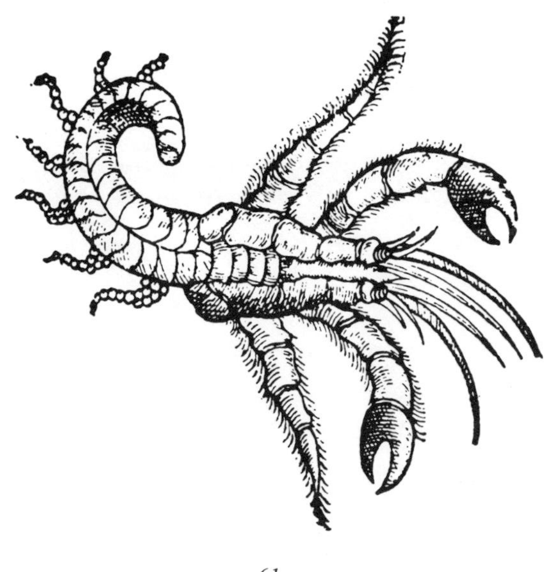

61c.

61. Diverse shells, together with the fish which is within them, called Bernard the Hermit

or, Pinne]; which is that species of large shell which one calls
pearl oyster" or "naker," always remaining seated like a doorman 181
t the opening of the latter, holding it half-open until it sees some
mall fish—of the kind it can easily catch—entering in; the pin-
thera biting the naker, it closes its shell; then the two nibble and
at their prey together.

Rondelet, in the 3 book of *On Fish,* chap. 11, writes that this sh, is sometimes found [to be] so miraculously large, that it can carcely be dragged on a cart by two horses. It eats (says he) other shes, and is very gluttonous; indeed it devours whole men: which as been found out through experience. For at Nice and at Mareilles dogfish have been caught in whose stomach an entire, comletely armed man has been found. [Fig. 62]

"I saw (says Rondelet) a dogfish in Saintonge, which had such large throat that a big fat man might easily have entered it: so

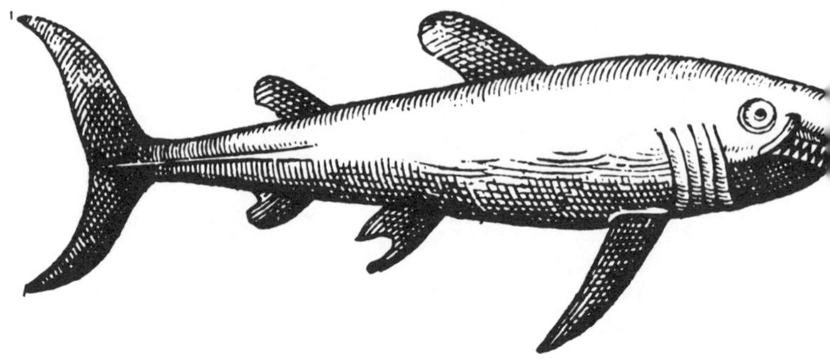

62. Figure of the Lamie

much so that if, with a gag, one holds the mouth open, dogs ente it without difficulty, in order to eat what they find in the stomach

Whoever might wish to know more concerning this may rea Rondelet in the specified place. Similarly, Conradus Gesnerus i his *History of Animals,* folio 151, order 10, confirms what Rondel has written about it; and he says besides that whole dogs have bee found in the stomach of said dogfish, having opened the latter, an that it has pointed teeth, sharp and large. Rondelet also says th they [i.e., the teeth] are of a triangular shape, jagged on both sid like a saw, arranged in six rows, the first of which is visible outsid the maw, and tending toward the front; those of the second ar straight; those of the third, fourth, fifth and sixth are curved towar the interior of the mouth for the most part. Goldsmiths decorat these teeth with silver, calling them snake's teeth. Women han them around their children's necks and think that they are ve beneficial to the children when they are cutting teeth; also that the keep them from being afraid.

I have a recollection of having seen, in Lyons, in the house o a rich merchant, a head of a large fish, which had teeth similar t this description, and I was unable to find out the name of this fis I believe now that it was the head of a dogfish. I had propose showing it to the now defunct King Charles, who was very intereste
182 in seeing large, weighty, and monstrous things; but two days aft

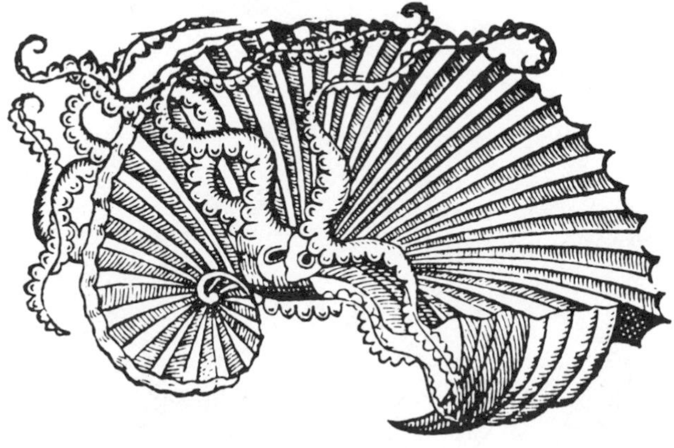

63. Figure of the fish called Nauticus, or Nautilus

I tried to get it brought to him, I was told that the merchant, his wife and two of his servants were stricken with the plague; which was the reason why he didn't see it.

Pliny, chap. 30, book 9 of his *Natural History*, names this fish *Nautilus* or *Nauticus*, which is very much to be considered, in that in order to come above the water, it turns about, mounting little by little to let the water that might be in its shell flow out, so as to make itself lighter to navigate, as if it had exhausted the sink of its ship. And being above water, it crooks two of its feet upward again, which [feet] are joined together with a thin film to serve it as a sail, using its arms as oars, even keeping its tail in the middle, in place of a rudder; and thus it goes over the sea, behaving like the logs and galleys. Should it feel that it is afraid, it folds in its equipment and fills its shell with water, and while diving and sinking its shell, it thus makes its way to the seabed.

([Next is the] figure of a *chancre*, or sea crab, that the Physicians and Surgeons have compared to a cancerous tumor, for the reason that it is round and rough and the veins surrounding [such a tumor are similar] to the twisted feet of this animal; also when it is hooked to rocks, it is detached with great difficulty; moreover, it is of a
dark [brown] and blackish color, as are cancerous tumors; and that 183
is why the Ancients gave the name of cancer to such a tumor,

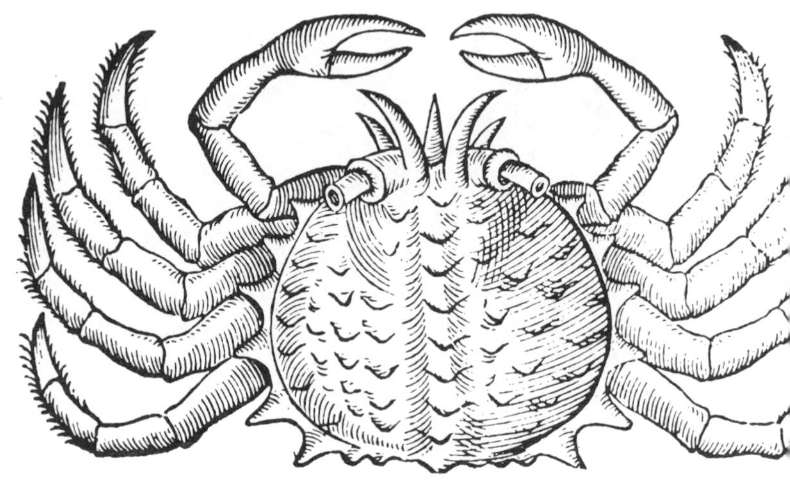

64. Figure of a Sea Crab

because of the similarity that they have to one another. *Chancres* [or crabs] are found in the hard shells of mussels and oysters, and other fish, which have shells, in order to be nourished and preserved in them, as within caves and fortified mansions, because there is no animal which does not have that gift of Nature of obtaining what is necessary to it, both for nourishing itself and for withdrawing and sheltering itself. Fishermen (says Aristotle) say that they are born with those in whose shells they are found. *Chancres* have ten feet, including their arms, [which are] forked and, in the interior, toothy, in order that they may use them as hands. Their tail is folded back on top; they are covered with sharp cockles, made in half-circles; they have six horns on the head, and eyes very much protruding and very separated from one another; in the spring they shed their shell, like a snake its skin, and feeling weakened and unarmed, they keep themselves hidden in the hollows of the rocks until their shell is replaced and hard. [1573])

Description of the Whale

We are stretching the word *Monster* somewhat for the greater enrichment of this treatise; we shall put the Whale in this category and we shall say it is the largest monster-fish that is found in the ea, most often thirty-six cubits long, eight wide; the opening of the mouth is eighteen feet, without [its] having any teeth; but instead [of teeth], on the sides of the jaws it has excrescences whalebones; baleen] as of black horn, which end in hairs similar o a swine's bristles, which come out of its mouth and serve it as a guide to show the way, so that the whale will not bump against he rocks. Its eyes are four ells apart, and [are] bigger than the head of a man; the snout [is] short, and in the middle of the forehead [there is] a conduit through which it draws in air and casts [out] a great deal of water—like a cloud—with which it can kill skiffs and other small vessels, and overturn them into the sea. When it is sated, it bellows and cries so loud that one can hear it for a French league; it has two large wings at its sides, with which it swims and hides its young when they are afraid, and on its back it has none at all; its tail is similar to that of a Dolphin, and, agitating it, it moves the water so greatly that it can overturn a skiff; it is covered with tough, black hide. It is certain by its anatomy that it gives birth to live little ones, and that it nurses them; for the male has testicles and a genital member, and the female has a womb and dugs.

It can be caught at certain times of winter in several places, even on the coast of Bayonne, near a small town—being at a distance of three leagues or thereabouts from said city—named Biarritz; to which I was sent by order of the King (who was then at Bayonne), to treat His Highness the Prince of La Roche-sur-Yon, who remained ill there; where I learned and confirmed the means that they use to do this, which I had read about in the book monsieur Rondelet has written about fish, which is as follows: Opposite said town there is a small mountain, on top of which—for a long time—a tower has been built expressly to keep watch, both by day and by night, to discover the whales that pass in this place; and they perceive them coming, both on account of the great noise they make and also on account of the water that they cast [up] through a conduit that they have in the middle of the forehead; and having perceived it coming, they ring a bell, at the sound of

65. Figure of a whale-catch

which all the inhabitants of the town run to the spot with whatever of their equipment is necessary to catch it. They have several vessels and cockboats, in some of which there are men strictly designated to fish out those [men] who might fall into the sea; the others [are] specified to do combat, and in each [boat] there are ten robust, powerful men who can row hard, and several others therein, with barbed darts [harpoons]—which are [each] marked with their [own personal] stamp, so they can recognize them—attached to ropes; and with all their might they throw them upon the whale, and when they perceive that it is wounded—which is recognized by the blood that is issuing from it—they loosen the ropes of their darts [harpoons], and follow it so as to fatigue it and catch it more easily; and drawing it on board, they rejoice and

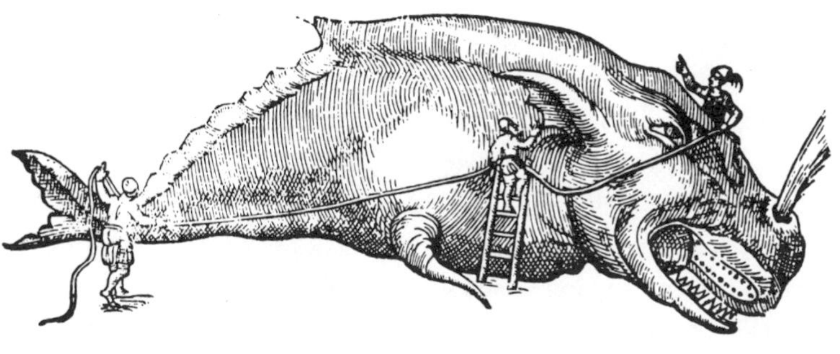

66. Another species of Whale [1573; 1579]

are merry; and they divide [it] up, each getting his portion according to the duty he will have performed; which is reckoned by the quantity of darts they will have thrown and that will have been found, which remain within [the whale]; and they recognize them by their [particular] mark. Now the females are easier to catch than the males, because they are concerned with saving their young and spend effort only in hiding them, and not in escaping.

The flesh is not in the least bit prized; but the tongue, because it is soft and delicious, they salt it [down]; similarly the bacon which they distribute into many provinces, which is eaten at Lent with peas; they keep the lard [or, fat] to burn, and to rub [on] their boats, which, being melted, never congeals. From the baleen that come out of the mouth, vardingales [staves] and bustles are made for women, and [also] knife handles, and several other things; and as for the bones, those [inhabitants] of the country make fences for gardens out of them; and from the vertebrae, footstools and seats for sitting on in their houses.

I had one of them brought [to me] that I keep in my house as [an example of] a monstrous thing.

[Above is a] true picture of one of the three whales that were caught the second of July 1577, in the River Scheldt, one at Flushing [Vlissengen], another at Saflinghe, and this one here at Hastings-
184 au-Doel, about five leagues from Antwerp; it was of a dark blue color; it had on its head a nostril [spout hole] through which it cast water; it had in all a length of fifty-eight feet and a height of sixteen feet; the tail [was] fourteen feet wide; from the eye as far as to the front of the snout there was an expanse of sixteen feet. The lower jaw was six feet long, on each side of which were twenty-five teeth. But on top it had as many holes in which said lower teeth could be hid. A monstrous thing, to see the upper jaw devoid of the teeth which ought to have been opposite [to the lower teeth] for encountering food, and in place of these teeth to see useless holes. The biggest of these teeth was six inches long, the entirety was wondrous and frightening to contemplate, due to the vastness, largeness, and thickness of such [an] animal.

Pliny, book 32, chap. 1, says that there is a wretched little fish
185 only half a foot big, named by some *Echeneis,* by others *Remora,* which very much deserves to be set here among miraculous and monstrous things, which detains and stops seagoing vessels—however big they may be—when it attaches itself to [them], no matter what effort the sea, or men, may put forth to the contrary, [such]

as surfs, and waves and the wind being engulfed by the sails, and backed by oars and cables, and anchors, however thick and heavy they may be. And, in fact, they say that at the defeat of Actium, 186
in the city of Albania, this fish stopped the Admiral Galleyship in which was Marcus Antonius, who, by dint of oars [i.e., row boats], went from galley to galley, giving courage to his people; and during this, the armada of Augustus, seeing this disorder, surrounded that of Marcus Antonius so abruptly that he crushed him. The same thing happened on the galley of Emperor Caligula. This Prince, seeing that his galley alone among all those of the armada wasn't moving ahead, and yet was [a galley] with five [rowers] to every bench, suddenly understood the reason why it was stopping: many divers immediately jumped into the sea to seek out—around this galley—what was making it stop, and they found this little fish attached to the rudder, the which being brought to Caligula, he was very angry that such a little fish had the power to set itself against the effort of four hundred front-row oarsmen and galley oarsmen who were in the galley.

(Moreover, Pliny, in the same book and chapter, says there is another fish named torpedo [or, cramp-fish] which by merely touching the line stupefies and deadens all feeling in the arm of the person who holds the line. [1575, 1579, omitted 1585]) Listen to this great wise Poet, Milord Du Bartas, who utters with such great grace in the fifth book or *La Sepmaine,* the following verses:

> The *Remora* fixing her feeble horne
> Into the tempest-beaten Vessel's sterne,
> Stayes her stone-still: while all her stout Consorts
> Saile thence at pleasure to their wished Ports.
> Then lose they all the sheates, but to no boot,
> For the charm'd Vessell bougeth not a foot;
> No more then if three fadome under ground
> A score of Anchors held her fastly bound:
> No more then doth an Oake, that in the Wood
> Hath thousand Tempests thousand times with-stood,
> Spreading as many massie rootes below
> As mightie Armes above the ground doo grow.

Oh *Stop-Ship* say, say how thou canst oppose
Thy selfe alone against so many foes?
O tell us where thou doo'st thine Anchors hide,
Whence thou resistest Sayles, Owers, Wind and Tide?
How on the sodaine canst thou curbe so short
A Ship whom all the Elements transport?
Whence is thine Engine and thy secret force
187 That frustrates Engines, and all force doth force?

36
ON FLYING MONSTERS

his bird is called an Ostrich, and is the largest of all, almost partaking of the nature of four-footed
188 animals, very common in Africa and Ethiopia; it does not budge from land and take to the air, and nevertheless it surpasses a horse for speed. It is a miracle of nature that this animal digests all things indifferently. Its eggs are of a miraculous size, to the point that one can make vases of them; its feathers are very beautiful, as one can recognize and see by this picture.

I do not want to fail to speak, either, of the rarity I saw concerning the bones of the Ostrich. The late King Charles, having
189 three of them kept at the home of monsieur le Mareschal de Rets, one of which having died, it was given to me, and I made a skeleton [i.e., skeletal diagram] of it. The portrait of which I wanted to insert here, [along] with its description. [Fig. 68]

A. The head is a little bit bigger than that of the crane, one
190 span long from the summit of the head extending to the beak, [the head] being flat, having a beak slit up to about the middle of the eye, this being a trifle round at its extremity.

B. Its neck is three feet long, composed of seventeen vertebrae, which have at each side a transverse apophysis [or process] extending downward, a good inch long, except that the first and second near the head do not have any, and are joined together by ginglymus.

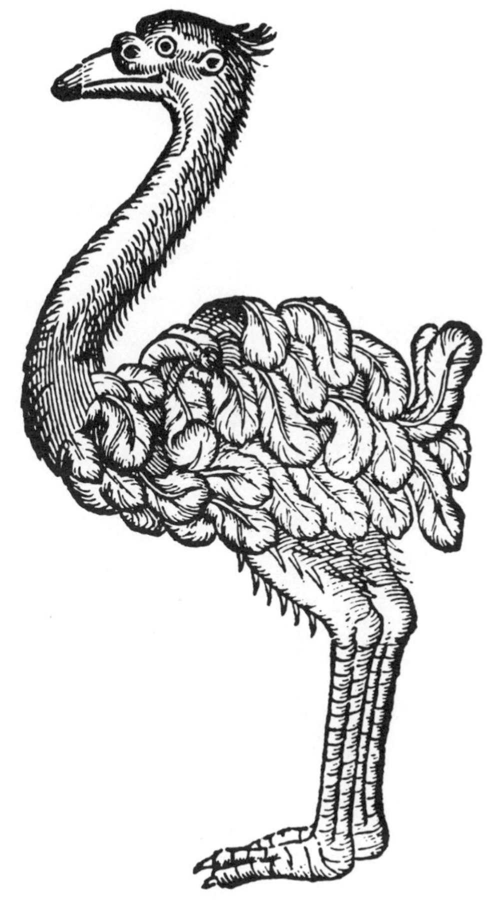

67. Figure of an Ostrich

C. Its back, one foot in length, is composed of seven vertebrae.

D. The Sacrum bone is two feet long, or thereabouts, on top of which there is a transverse apophysis beneath which there is a large hole, E, then three other smaller ones, F, G, H; following which there is a box into which the hip bone is insinuated, I, producing on its lateral external part a pierced bone, K, almost at its beginning, then, [it] is united; afterward, said bone forks in two [or, bifurcates], one of which is thicker, L, and the other is lesser,

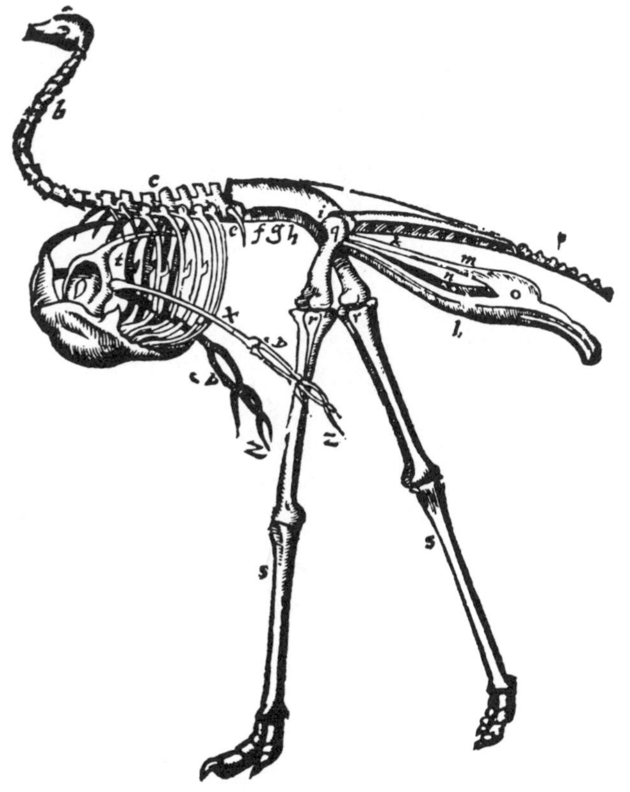

68. Portrait of an Ostrich skeleton

191 M, each one [being] a half-foot four fingers long: then they reunite having—between the place where they fork and where they reunite—a hole four fingers wide, N, and longer than a span: then what bone remains is in the shape of a bush-hook or a crooked knife, wide by three fingers breadth, and six inches long, O; then at its extremity it is joined by synchondrosis.

P. The tail bone has nine vertebrae, similar to those of man.

There are two bones in the thigh, the first of which, Q, the thighbone, is a good foot in length and as thick as that of a horse or thicker; R, the other, which follows it, is a foot and a half in ength, having on top a small shank of the length of the bone, osing its pointedness toward the bottom.

S. The leg to which the foot is attached is a foot and a half ong, having at its extremity two claws [nails, ungues], one large nd the other small; there are three bones to each claw [unguis].

T. Eight ribs which are inserted into the bone of the Sternum, rom which—in the three in the middle, on each side—there is bony extension resembling a hook.

V. The bone of the Sternum is of a piece, a foot larger, rep-esenting a shield [or, buckler] to which is joined a bone that rides on] the first three ribs, which takes the place of clavicles.

X. The first bone of the wing is a foot and a half long.

Y. Above it there are two other bones resembling the Radius nd Cubitus, at the end of which are attached six bones, Z, which re at the extremity of the wing.

The entire animal is seven feet in length and seven feet and nore in height, starting at the beak and ending at the feet.

There are several other remarkable things that I am setting aside or the sake of brevity.

Thevet, in his *Cosmography* (Book 21, Chap. 12), says he has een in the newly discovered lands a bird that the savages call in heir speech *Toucan,* which is very monstrous and deformed, in-asmuch as it has a beak [that is] thicker and longer than all the est of the body. It lives on pepper [corns], as our turtle doves, hrushes, and starlings do here on ivy berries, which are no less ot than pepper.

A Provençal gentleman made a present of one of them to the ate King Charles IX, which he could not manage to do [with the bird still] alive, for while taking it to him, it died; nevertheless he presented it to the King, who, after having seen it, ordered milord he Mareschal de Retz to hand it over to me, to anatomize it and o embalm it, in order to preserve it better; even so, soon afterward, t became putrefied. It was in thickness and in plumage similar to a Crow, leaving aside [the fact] that the beak was larger than the rest of the body, of a transparent, yellowish color, very light, and

69. Figure of the bird named Toucan

having teeth like a saw. I keep it as [an example of] an almos monstrous thing. The figure of which is represented for you here

Jerome Cardan, in his book *De Subtilitate* [*rerum*], says that i the Moluccas [or, Spice Islands], one can find on land or on sea a dead bird called *Manucodiata*, which in the Indic language signifie "bird of God," which one never sees alive. It lives high up in th air, its beak and body similar to the swallow, but adorned wit diverse feathers: those which are on top of the head are simila to pure gold, and those at its throat to those of a duck; its tail an wings similar to those of a peahen. It has no foot, and if lassitud overtakes it, or else should it wish to sleep, it suspends itself b its feathers, which it twists around the branch of some tree. Thi [bird] flies at a miraculous speed and is nourished only by air an

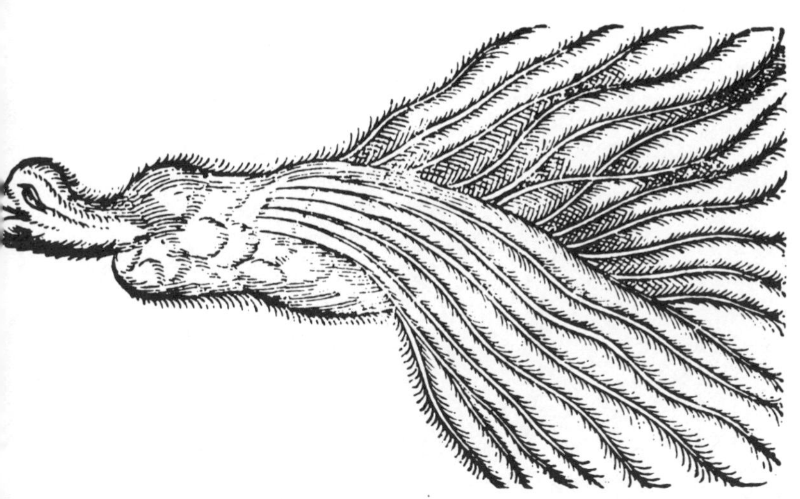

dew. The male has a cavity on its back, in which the female broods its young. (The interior of this bird, as Melchior Guillaudin Beruce describes, is stuffed and replete with fat, and he says he has seen two of them. As for me, I have seen one of them in this city, that a noteworthy man had, [and] that he held in high esteem: the picture of which bird you have here. [1573, 1575])

I have seen one of them, in the city, which was given to the late Charles IX; and also I keep one of them in my office, which [bird] was given to me on account of its pre-eminence. 192

37

ON TERRESTRIAL MONSTERS

André Thevet, vol. 1. book 4. chap. 11, says that on the Island of Zocotera, that one sees an animal 193
which is called Huspalim, as large as an Ethiopian 194

71. Figure of an animal called Huspalim

marmot, very monstrous—which the Ethiopians keep in great rush cages—having skin as red as scarlet, a little bit spotted, its head [as] round as a ball, its feet round and flat, without harmful claws [nails; ungues], which lives only on wind. The Moors club it to death, then eat it, after having beaten it several times, in order to render its flesh more delicate and easy to digest.

In the kingdom of Camota, of Ahob, of Bengal, and other
195 mountains of Gangapur, Plimatic, and Caragan, which are in interior
India, beyond the Ganges River, some five degrees beyond the
Tropic of Cancer, is found the animal called, by the Western
Germans, the Giraffe. This animal differs little in its head and ears,
196 and in its cloven feet, from our Hinds. Its neck is about one *toise*

72. Figure of the Giraffe

long, and marvelously slender, and it differs likewise in its legs, in that his are as highly elevated as [those of] any animal that may exist under Heaven. His tail is round, but does not reach further than his hams; his skin the most beautiful possible. It is spotted in several places, these spots alternating between white and tawny, like those of the Leopard, which has given to some Greek Historiographers the argument that he should be given the name of Chamoeleopardalis. This animal is so wild in the face of being caught, that very seldom does it allow itself to be seen, hiding in the woods and deserted places of the country, where other animals do not at all feed; and just as soon as it sees a man, it tries to run away; but one can finally catch it, because it is slow in its course. Moreover, once it is caught, it is the gentlest animal to govern of any other in existence. On its head appear two small horns, a foot

long, or thereabouts, which are rather straight and surrounded by hair all about; a lance is not taller [than it] when it raises its head up high. It grazes on grasses, and lives also on leaves and branches of trees, and it likes bread very much, a thing that André Thevet attests to and pictures, book 11, chap. 13, volume 1, of his *Cos-*
197 *mography.*

Elephants are born in Africa, beyond the deserts, in Mauretania, and also in Ethiopia. The largest are those that are born in India. They surpass in largeness all other four-footed animals; nevertheless, as Aristotle says, they can be so readily tamed that they remain the most gentle and tractable of all beasts; one can teach them, and they understand how to carry out several charges. They are covered with a hide similar to a wild ox, covered with hair of an ashen color. They have a thick head, short collar, ears as broad as two spans, a very long and hollow nose, like a great trumpet,

73. Figure of the Elephant

almost touching the ground, which they use in place of hands. They have their mouth near their chest, rather similar to that of a swine; from the top issue two very large teeth. Their feet are round like trencher plates, two or three spans wide, and around them are five nails [ungues]. They have thick, strong legs, not composed of a single whole bone, as some have guessed, but [rather] they have knees that bend like other four-footed beasts; and as a result, when one wants to climb onto [one of them] or load them, they kneel down and then they get up again. They have a tail like a water buffalo [or, wild ox], not furnished with much hair, and [only] about three spans long; as a result of which they would be badly used by flies, if Nature had not provided them with another means of defending themselves; [and] that is, that when they bite and sting them, they contract their hide, which is all wrinkled and full of folds; thus they catch and crush them among their wrinkles. There's no man that he can't outrun, while still going at his own pace; his great corpulence is the reason for this; for his steps are so long that they overtake the great speed of man. They live on fruit and leaves from trees, and also there is no tree so large that they cannot uproot it and dash it to pieces. They grow to the height of sixteen spans; for this reason, people who aren't accustomed to traveling atop them are as aghast as those who haven't the habit of traveling on the sea. They are so unruly in their nature that they cannot endure any kind of a bridle whatsoever, which is the reason that one must let them go at liberty; nonetheless they are very obedient to the men of their nation, understanding their language very well; whereby it is easy to govern them through words. When they want to molest some person, they raise him in the air with their great nose [or, trunk], then in a burning fury they fling him to the ground and trample him under their feet, until they have made him give up the ghost.

Aristotle (Book 6, Chap. 27, of *Hist. animal.*) says that they don't reproduce before twenty years of age; they are not at all adulterous, for they never touch but one female, and when they know her to be pregnant, they have no mind to touch her. One cannot find out how long the female carries her young, for the males cover her in secret, for the shame they have. The females deliver their young with pain like women, and they immediately lick them. They see and walk just as soon as they are born. They live two hundred years.

One sees Elephant teeth [tusks], called Ivory, wondrously large, in several cities of Italy, such as in Venice, Rome, Naples, and even in this city of Paris, from which one makes chests, lutes, and combs,
198 and many other things for man's use.

André Thevet, vol. 1, chap. 10, in his *Cosmography,* says that at the time he was on the Red Sea, certain Indians arrived from *terra firma* who brought [with them] a monster of the size and proportion of a Tiger, having no tail, but its face completely similar to that of a well-formed man, and the hindfeet resembling those of a Tiger, completely covered with tawny to black body hair. And

74. Figure of the animal Thanacth

as for the head, ears, neck, and mouth: like a man, having head hair slightly black and kinky just like the Moors whom one sees in Africa. That was the novelty that these Indians were bringing to show, through the kindness and courtesy of their land, and they called this gentle beast *Thanacth;* which [animal] they kill with arrow wounds; then they eat it.

Thevet, in his *Cosmography,* vol. 2, chap. 13, says that in Africa is found a very deformed animal, called by the Savages Haiit; and it is almost unbelievable to him who has never seen it that there should be such a one. It can be in size as big as a large Guenon, having its belly hanging down and close to the ground, even though it be standing; its face and hands are almost like those of a child.

75. Figure of a monstrous animal which lives only on air, called Haiit

This Haiït being caught, it heaves great sighs, more or less as a man afflicted by some great and excessive pain would do. It is a gray color, having only three claws [ungues] to each paw—four fingers long—[the claws] made in the shape of the bones of a carp, with which claws, which are as trenchant as—or more trenchant than—those of a Lion, or some other cruel beast, it climbs up into the trees, where it makes its dwelling more than on the ground. It has a tail only three fingers long. Moreover, it is a strange case, for never could any man say he has seen it eat anything whatsoever, although the Savages have kept some of them in their huts for a long time in order to see if they would eat something; and the Savages say that they live on wind.

199 I have taken from John Leo, in his *African History,* this very monstrous animal, round in form, similar to the Tortoise; and on the back, there are two lines crossing each other at right angles, in the form of a cross, [and] at the end of every line is an eye and

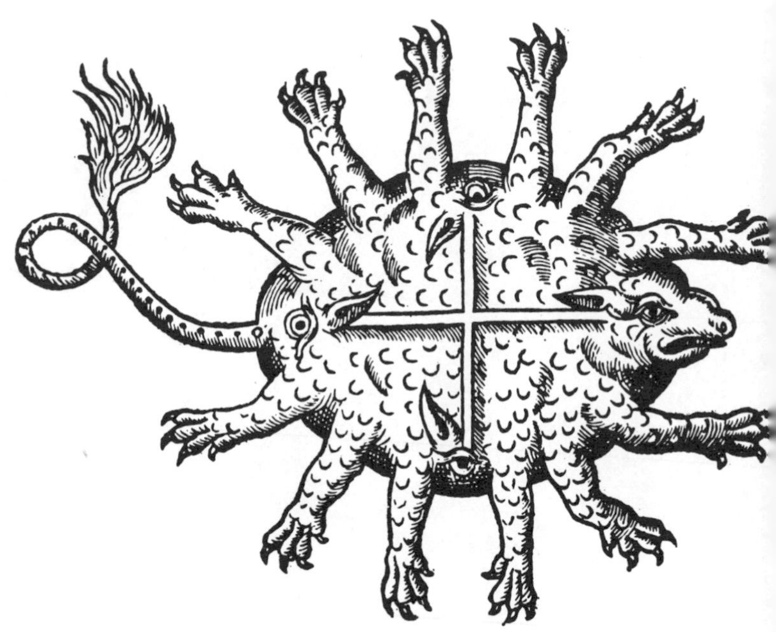

76. Figure of a very monstrous animal that is born in Africa

an ear, so that in four directions and on all sides these animals see and hear, with their four eyes and with their four ears, and yet they have only one single mouth and belly, to which what they drink and eat descends. These animals have several feet [distributed] around their body, with which they can make their way in any direction they wish without turning their bodies around; the tail is rather long, the end of which is very thickly tufted with hair. And the inhabitants of this country affirm that the blood of this animal has miraculous virtue in knitting and closing up wounds, and there is no balm which has greater power for doing this.

But what person will not marvel greatly, upon contemplating this animal, having so many eyes, ears and feet, and [upon seeing] each one do its office? Where can the instruments dedicated to such operations be? Truthfully, as for myself, I lose my mind over it and could not say anything other than that Nature has disported herself [played a trick] in order to cause the grandeur of her works to be admired. 200

One finds this animal, named *Chameleon,* in Africa, and it is made like a lizard, except that it is taller in the legs; moreover, it has its flanks and belly [joined] together, like a fish; it therefore

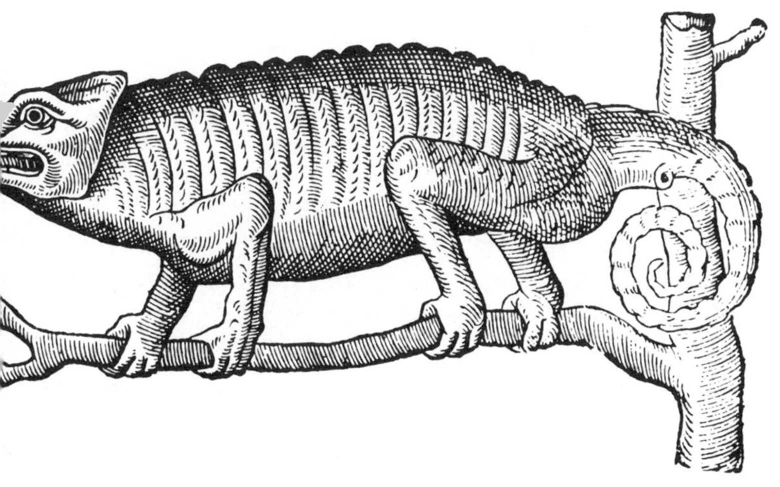

77. Figure of the Chameleon

has a ridged spine over the back, as one sees in fish: it has a snout like a small pig, a very long tail which narrows down to a point [at the end], its nails [ungues] [are] very sharp, and it walks as heavily as a Tortoise, and has a body [that is] rough and scaled like a Crocodile; it never closes its eye, and does not budge its eyelid at all. Besides, it is a wondrous thing to speak of its color: for at all hours, chiefly when it inflates itself, it changes it; which occurs for the reason that its hide is very fine and thin, and its body transparent; so that one of two things occurs: either that in the tenuity of its transparent hide the color of the things that are nearby it is easily reflected, as in a mirror (which is the more likely), or that the humors diversely stirred up in it, according to the diversity of its imaginings, represent diverse colors toward the hide, not differently from the excrescences [or, wattles] of a Turkey
201 cock. Being dead, he is very pale. Matthiole says that if one tears its right eye out while it is alive, it can clean away white spots which are upon one's cornea, [when] mixed with some goat's milk; if one rubs oneself with its body, [one's] body hair falls [off]; its gall
202 discusses [scatters] and lifts cataracts from the eyes.

I have observed this description in the one I have in my home.

38

ON CELESTIAL MONSTERS

The Ancients have left it to us in writing that the face of the Sky has so many times been disfigured by bearded [and] hairy [i.e., blazing] Comets, by torches, lamps, pillars, lances, bucklers, battalions of clouds, dragons, duplication of Moons and Suns, and other things; which I did not want to omit, in order to make this book on Monsters complete; and for this reason, I shall first of all produce this story, which figures in the *Histoires prodigieuses* of Boistuau, who, says he, took
203 it from Lycosthenes.

Antiquity, says he, has experienced nothing [that is] more to be marvelled at in the air than the horrible blood-covered comet which
appeared in Westrie, on the ninth day of October 1528. This 204
Comet was so horrible and frightening, that it engendered such great terror in the common people that some of them died of fear over it; the others fell ill. This strange comet lasted an hour and a quarter, and began to appear in the direction of the rising Sun,

78. The figure of a fearful comet seen in the air

then drew near the Midi [southern France]; it seemed to be of an excessive length, and so it was of the color of blood; at the summit of this [comet] one could see the shape of a curved arm holding a large sword in its hand, as if it would have liked to strike. At the end of the point there were three stars, but the one that was directly on the point was more bright and shiny than the others. On the two sides of the rays of this Comet there could be seen a great number of hatchets, knives, swords, colored with blood, among which there was a great number of hideous human faces, with their beards and hair bristling.

205 Josephus and Eusebius wrote that after the Passion of Jesus Christ, the wretched destruction of the city of Jerusalem was declared by several signs, and even, among others, a frightful comet in the form of a shining sword on fire, which appeared for easily the space of a year on the temple; as [if] demonstrating that divine ire wished to take vengeance over the Judaic nations, through fire, through blood, and through famine. Which [indeed] did happen, and there was such a calamitous famine that the mothers ate their
206 own children; and there perished in that city, at the siege by the Romans, more than twelve hundred thousand Jews; and more than ninety thousand of them were sold.

Comets have never appeared without producing some bad effect, and leaving [behind] a sinister outcome. The poet Claudian [writes]:

> Never has anyone seen a Comet in the sky
> 207 But that some harm it brought about for us.

The astronomers have divided the celestial bodies into two groups; one called fixed or arrested stars, which one can see sparkling and twinkling in the Sky, as if they were hot coals; others are drifting [and are] called planets, which do not sparkle at all and there are seven of them in number, each one having its sky circle, orb, or storey; their names are Saturn, Jupiter, Mars, Sol Venus, Mercury and Moon. Stars are spheric bodies, that appear and shine, comprised of simple and pure matter, like the Sky, and no one knows the number of them, nor their names, except God Now the aforementioned planets make their course by the Zodiac

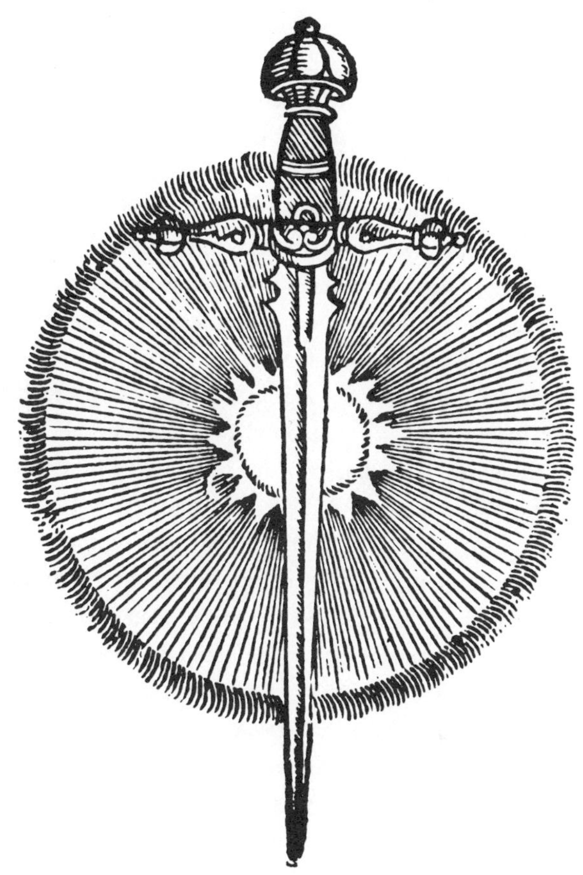

79. Figure of a Comet

which is one of the chief [circles] and [is also] the largest circle f the Sky and the true route of the Sun), which crosses or ncompasses the Sky obliquely, night and day, so that all the counries of the earth may enjoy alternately four seasons of the year, y means of the Sun which ceaselessly rises and sets, lighting and eeding the whole orb of the earth in the space of one year. It is he chariot and the fountain of light of the celestial bodies, [they] eing only small streams of it; wherefore it is named the King of he stars, and the greatest of all the celestial bodies. It is at three picycles, that is to say, skies, or storeys, above the Moon; it proceeds in the middle of six planets; if they approach it, in orde not to impede its course they pull aside to the highest [spot] i their small epicycles or circles; then, once it has passed by, the set, at the lowest point, in order to accompany and flank it, a princes do their King. And then, having done their duty, they stop and with a bashful bow, they draw backward, descending to th bottom of their epicycles, in order to contemplate, as from afa the face of their "lord." And when it draws near, pulling back, the regain the highest [spot] in their epicycles, in order to go fort to meet it; with the result that, feeling it [to be] as near [to them as four [celestial] signs [of the Zodiac] away, they pretend to wa for it, then, having bid it a welcome, they proceed ahead of it— [and] slightly apart [from it], in order not to give hindrance to i natural scope and course.

The one which is named Saturn, according to the estimate o the astronomers, is approximately eighty times larger than the whol earth, from which it is removed by more than thirty-six millio French leagues. The size of the one named Jupiter is estimate [to be] ninety-six times larger than the diameter of the earth an is removed from it by more than twenty-two million leagues. Th planet of Mars is as large as the earth and is removed from it b three million fifty four thousand two hundred and four league *Moon* signifies months, because every month it renews itself; it removed from the earth by eighty thousand two hundred thirtee leagues; it is thicker and darker than the other stars, [is] attache to her sphere, which carries her through certain movements an journeys to and fro, [these latter] being limited; [it was] create by God to mark out for men the times and the seasons, and t work by its light and movement on inferior bodies.

The globe of the Sun is sixty-six times larger than that of th earth, and is almost seven thousand times larger than the Moo Ptolemy and other astronomers have found through geometr calculations that it was one hundred sixty-six times larger than th whole earth; it gives life to all animals, not only those which a on land, but also those which are in the depths of the waters. Lo Du Bartas calls it *continual postillion, fountain of heat, source of ligh*

fe of the universe, torch of the world, and *adornment of Heaven.*
Moreover, the Sun makes its tour of the Sky around the earth in
wenty-four hours, and causes the benefits and [the] agreeable
evolutions of the day and of the night, for the relief and content
f man, and of the animals. 208

Let the reader consider and here adore the awesome wisdom
nd power of the Creator, in the size, continual speed, unbelievable
apidity, immense radiance and warmth, and [in the] conjunctions
nd opposite movements [found] in such a noble body as that of 209
e Sun, which in one single minute out of an hour goes several
ousands of leagues without our seeing it budge; and one rec-
gnizes nothing of this until after it is very far advanced on its
ourse. What is more, the least star is eighteen times larger than
e whole earth. Let this be said not only [as a matter] for con-
derable contemplation, but to the praise of the Creator, and in
rder to humble man, who makes so much noise on earth, who is
othing but a pinpoint in the eyes of the celestial machine. 210

Moreover, there are in the Heaven twelve signs, to wit: Aries,
aurus, Gemini, Cancer, Leo, Virgo, Libra, Scorpius, Sagittarius,
apricornus, Aquarius, Pisces, all of which are different. The use
f these is that through their conjunction with the Sun, they augment
r diminish the heat of the latter, so that by such a variety of heat
e four seasons of the year may be produced, [and] life and
reservation may be given to all things. The skies are a quintessence
f the four elements made of nothing, that is to say, without matter.

Whooaa . . . my pen, stop! For I do not want to, nor can I, go 211
y further ahead into the sacred cabinet [office] of God's divine
ajesty. He who would wish to know more about it, let him read
tolemy, Pliny, Aristotle, Milichius, Cardan, and other astronomers,
d chiefly Lord Du Bartas and his interpreter who have written 212
such a learned way and with such inspiration about it in the
Fourth Day" of *La Sepmaine* [*The Week*], where one will find
nough concerning it] to be wholly contented; and I confess having
ken from there the things mentioned above, for the purpose of
structing the young Surgeon in the contemplation of celestial

matters. And here shall we sing with that great divine prophet
213 [David] Psalm 19.

> The heavens shew forth the glory of God, and the firmament declareth the work of his hands.

And in Psalm 8 [verses 3–4]:

> For I will behold thy heavens, the works of thy fingers; the moon and the stars which thou has founded. What is man that thou are mindful of him? or the son of man that thou visitest him?

Moreover, I do not want to fail here to write of things both
monstrous and wondrous which are created in heaven. And first
214 of all, Boistuau writes in his *Histoires prodigieuses,* that in Sugolie
situated on the border of Hungary, there fell a stone from heaven
with a horrible bursting noise, on the seventh day of September
1514, weighing two hundred fifty pounds, which [stone] the citizen
have had set [or locked] in a thick iron chain, [hanging] in the
middle of their temple, and it is shown with great marvel to those
who travel through their province, a miraculous thing how air can
sustain such weight.

Pliny writes that during the Cimbrian wars, there were heard
from the air sounds of trumpets and clarions, along with a great
clattering of weapons. He also says, moreover, that during the
consulate of Marius, there appeared armies in the sky, some of
which came from the East, others from the West, and they combated
against each other for a long while, and that those of the East
pushed back those of the West. This same [thing] was seen in the
215 year 1535 in Lusalia, near a burg named Juben, around two o'cloc
216 in the afternoon. Moreover, in the year 1550, on the 19 of July
in the country of Saxony, not very far from the city of Wittenberg
there was seen in the air a large stag (Chap. 17 [in the book c
Boaistuau]), surrounded by two huge armies, which made a grea
noise while combating each other, and at that very instant, bloo
fell on the ground, like a heavy rain, and the sun split into tw
pieces, one of which seemed to have fallen to the ground. Also

›efore the capture [fall] of Constantinople, there appeared in the .ir a large army with an infinity of dogs and other animals.

Julius Obsequens says that in the year 458, in Italy, flesh rained lown in big and little chunks, which was in part devoured by the ›irds of the sky before it might fall to earth, and the rest, which ell to earth, remained for a long time without spoiling or changing olor or odor. And what is more, in the year 989, [while] Otto :mperor III of that name [was] reigning, wheat rained down from he sky. In Italy in the year 180, it rained milk and oil in great bundance, and the fruit trees bore wheat. Lycosthenes tells that ı Saxony it rained fishes in great numbers; and that in the time f Louis [the] Emperor, it rained blood for three whole days and ights; and that in the year 989, snow as red as blood fell near ıe city of Venice; and that in the year 1565, in the Bishopric of)ole, it rained blood in a great quantity. Which happened in the ame year, in the month of June, in England. 217

And not only do monstrous things occur in the air, but also in ıe Sun and in the Moon. Lycosthenes writes that during the siege f Magdeburg, in the times of Emperor Charles V, right around even o'clock in the morning, three suns appeared, the middle one f which was very bright, and the others tended toward red and lood color, and they were visible all day; also three Moons appeared t nighttide. This same [thing] happened in Bavaria, 1554.

And if such news is engendered in the sky, we will find the earth roducing just as wondrous and dangerous effects, or more so. In ıe year 542 the whole earth trembled, and even Mount Etna omited many flames and sparks, because of which the greatest ıare of cities, and towns and properties of said Isle were consumed / fire.

(Moreover, in the year 1531 in Portugal it happened that the ırth trembled during eight days, and on each day seven or eight mes, so much so that in the city of Lisbon alone 1,050 houses ere leveled, not counting six hundred which were smashed and ıved in; and not long ago the city of Ferrara was almost ruined a similar quake (in the year 1551). Pliny recounts and says that his time, under the reign of Nero, that Vasseus Marcellus, a 218

Roman officer, had, in the territory of Marrucin, some fields o
both sides of the highway, the one being a meadow, the othe
planted with olive trees. It came about by a miraculous force th
219 these two fields changed places; for the olive trees were transporte
to the place where the meadow was, and the meadow, in a simil
way, was seen to be transported to the place where the olive tre
were, which was judged to proceed from an earthquake.
220 [1579])

39

[ON NATURAL DISASTERS]

braham Ortelius, in the *Théâtre de l'Univer* describes that there is in Sicily a burning mounta named Etna; several philosophers and poets have wri ten of this mountain, because it continually casts [out] fire ar smoke, which [mountain] is more than thirty Italian leagues hig and more than one hundred leagues around at the bottom, accordir to what Facellus writes, who has looked at it very carefully ar described it with no less scrupulousness. Above this continual flam that is never extinguished, it sometimes throws up such a quanti of fire that the whole countryside surrounding it is complete spoiled and burned. But how many times that has come about, o predecessors have not made a record of; nevertheless, what t authors *have* written about it, we will recount here briefly, ar according to what Facelle says.

221 In the year of the founding of the city of Rome, 350 [B.C.], th mountain vomited [up] so much fire that, by the embers and h coals that issued from it, many fields and towns were burned; 2 years afterward the same thing happened; 37 years after this, disgorged and cast [forth] so many hot ashes, that the roofs ar coverings of houses in the city of Catania, located at the foot this mountain, were ruined from the weight. It [i.e., the mountai similarly did great damage in the time of the Emperor Caligu

ıd then afterward in the year 254. The first day of February, in
e year 1169, it brought down, by the continual fire that issued 222
om it, several rocks, and caused such an earthquake that the reat Church of the city of Catania was demolished and razed; and e Bishop, along with the Priests and the people who were there the time, were struck down and crushed. In the year 1329, the st day of July, having made a new opening, it razed and ruined—th its flames and the earthquake which resulted from it—several hurches and houses located around said mountain: it caused veral fountains to dry up; cast several boats that were in dry dock to the sea, and at the same instant broke open again in three aces, with such violence that it overturned and tossed into the r several rocks, indeed, also, forests and valleys, casting up and omiting such fire through these four infernal conduits that it [the e] flowed down from said mountain to the bottom, like roaring reams, ruining and razing everything it encountered or that gave any resistance; the whole surrounding countryside was covered th ashes that had issued forth from those aforementioned burning aws at the summit of the mountain; and many people were suf-cated; in [such] a way that said ashes from that sulphured odor ere transported by the wind (which was then blowing from the orth) clear to the Isle of Malta, which is 160 Italian leagues moved from that mountain. In the year 1444, it rose up again, ry terribly, vomiting fires and pebbles. After that time, it ceased cast fire and smoke, so that people believed it to be totally tinct, and not due to erupt anymore. But that fair season (so to eak) was soon past. For in the year 1536, the 22 of March, it ain began to vomit many burning flames, which razed everything ey encountered in their path. The Church of St. Leo, located thin the forest, collapsed because of the quaking of the mountain, d immediately afterward it was so reduced to cinders by the fire at not one bit of it remains, except a pile of burned stones.

All this was a very horrible thing. But it was still not anything comparison with what happened since, in the year 1537, on the st day of May. First of all, the whole Island of Sicily shook during elve days; afterward a horrible thunder was heard, with a roaring burst, just like heavy artillery, as a result of which many house were reduced to nothing throughout the whole Island. This laste about the space of eleven days; after that it split in several an diverse spots, from which cracks and crevices there issued suc a quantity of flames of fire—which came down from said mountain—that in the space of four days they ruined and reduced t ashes all that there was for fifteen leagues around; indeed, also several villages were entirely burned and ruined. The inhabitant of Catania, and many others, abandoning their cities, fled to th fields. A little while afterward, the hole which is at the summit c the mountain, cast forth on three consecutive days such a quantit of ashes that not only was this mountain covered with them, bu what is more, it spread and was driven by the wind clear to th extremities of this island, indeed, beyond the sea, clear to Calabri Certain ships sailing on the sea with the intention of going fro Messina to Venice, three hundred Italian leagues away from th island, were spattered by the above-mentioned ashes.

Here is what Facellus writes about it in the Latin tongue—
223 [taken] from his tragic stories—but much more lengthily. Abou three years ago news reached Antwerp that said mountain ha greatly damaged the countryside with its fire. On that island [i.e Sicily] there were formerly several magnificent cities, such as Syr-
224 acuse, Agrigento, and others; at present Messina [and] Palerm are its principal ones.

Marco Polo, the Venetian, in the 2 book on the Oriental coun-
225 tries, chap. 64, says that the city of Quinsay is the largest city the world, and that it is one hundred Italian miles in circumferen in which [city] there are twelve thousand stone bridges, benea which vessels with high masts can pass. It is on the sea, like Venic He affirms having sojourned there; which I have gathered fro
226 the interpreter of Saluste Du Bartas, in his "Fourth day" of *Sepmaine,* folio one hundred sixty-six.

Awesome things likewise happen on water. For huge flames fire [spreading] across the water have been seen issuing from t abysses and whirlpools of the sea—a very monstrous thing— if the great quantity of water could not stifle the fire; in this G

shows himself incomprehensible, as in all his works. (Moreover, the waters overflowed so strangely and so miraculously that in the year 1530 the sea rose up so in Holland and in Zeeland [the Netherlands] that the whole island thought it would be drowned, and all the cities and towns were made navigable for a long space of time. Also in Rome, the Tiber overflowed with such violence that it submerged a great part of the city, to such a degree that in some streets the water surpassed the height of thirty-six feet. And even in these years just past, the Rhone overflowed in such a way that it overturned a part of the bridge of Lyons and several houses of La Guillautière. [1579]) Lucio Maggio in his *Discourse on earthquakes* says that it has been observed that due to an earth- 227
quake the sea water heated up in such a way that it caused the melting of all the pitch around the ships which were at the time in its radius, even to seeing the fish swimming on the water almost totally cooked, and countless numbers of persons and animals died because of the extreme heat. Similarly ships have been seen to cave in on calm seas in one moment, because they are passing over some abysses, where the water is dead still and powerless to sustain any burden. Moreover, in the sea there are rocks [made] of lodestone so that if the ships pass too near, because of the iron [in the ships], they are swallowed up and lost in the deepest [depths] of the sea. In sum, strange and monstrous things are found in the sea, which is proved by that great Prophet David, who says, [in] Psalm 104:

> "There on this sea the ships shall go. And also this sea
> dragon which thou hast formed to play therein." 228

APPENDIX 1

Items from the DISCOURSE ON THE UNICORN

The items in this appendix were located in the treatise on monsters in early editions of Paré's collected treatises; in the 1582 (Latin) edition of the collected works they were moved to the *Discourse on the Unicorn,* from which they have been extracted and reconstructed here, in the order they would have in the text. The chapter numbers in brackets refer to the present location in the *Discours de la Licorne* (Malgaigne edition of Paré's *Oeuvres complètes,* vol. 3, pp. 491–519). These same items are to be found in the Céard edition of *Des Monstres* as variants found in early editions of Paré's work.

A. [Chapter 14] Fish resembling the Wild Boar-Hog in the head

Gesnerus says that in the "Ocean sea" a fish is born having the 229
head of a "wild boar-hog," which is of miraculous size, being covered with scales placed [on him] through the great order of Nature, having very long canine teeth, trenchant and sharp, similar to those of the "wild boar-hog." [1579; Fig. 80]

B. [Chapter 11] Sea Elephant

Hector Boetius, in the book that he wrote on the description 230
of Scotland, says that the animal whose likeness follows hereafter, is named "Sea Elephant," and [it is] larger than an elephant; which [Sea Elephant] lives in water and on land, having two teeth similar to those of an elephant, by which—when he wishes to take his sleep—he attaches himself and hangs onto the rocks, and he sleeps so soundly that when sailors catch sight of him, they have the leisure to come ashore and to tie him with thick ropes in several spots. Then they make a great noise and throw stones at him to

80. Figure of the Marine Boar

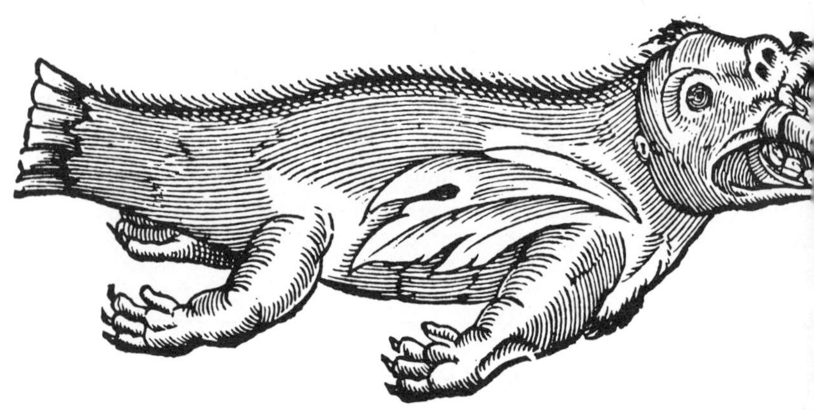

81. Figure of a Sea Elephant

wake him up; and then he tries to cast himself with great violence—as is his custom—into the sea. But seeing himself caught, he becomes so peaceful that one can easily take possession of him; they bludgeon him and pull off the fat; then they skin him, in order to make leather thongs and belts of [his hide], which—because they are strong and do not rot—are highly prized. [1579]

C. [Chapter 10] Pyrassoupi, Species of Arabian Unicorn

As one goes along the coast of Arabia on the Red Sea, one discovers the island named Cademoth by the Arabs, on which, towards the district that runs along the Plate River, is found an 231
animal that the savages call Pyrassoupi, as large as a mule, and his head almost identical [to the mule], his whole body shaggy in the form of a bear, a little more [highly] colored, tending toward yellowish, or dun, having cloven feet like a stag. This animal has two horns on its head—very long, without antlers, highly elevated, which approximate [those of] unicorns; which the savages [natives] use when they are wounded or bitten by venom-bearing animals, putting them in water for the space of six or seven hours; then

82. Figure of the Pyrassoupi

afterward they have the patient drink said water. (And its picture is taken from the fifth book of the *Cosmography* of André Thevet.) [1579]

D. [Chapter 4] About the Camphurch, an amphibious animal

André Thevet, in his *Cosmography,* says that another one [of these horned animals] is found in Ethiopia, almost identical, called Camphurch on the Island of Molucca, which is amphibious, that is to say, living in the water and on land, like a crocodile. This animal is of the size of a hind, having one horn on the forehead, mobile, three and a half feet long, as thick as the arm of a man,

83. Figure of the Camphurch

full of hair around the neck, tending toward a grayish color. It has two paws like those of a goose, which it uses to swim, and the other two forefeet like those of a stag or a hind; and it lives on fish. There are some [persons] who are persuaded that it was a species of Unicorn, and that its horn is very powerful and excellent [as a remedy] against poisons.

(The King of the island gladly bears the name of this animal, just as the other lords—among the greatest after the King—take the name of some other animal, some of fish, others of fruits, as André Thevet has depicted and described in his *Cosmography* (Book 12, chapter 5, volume 1). [1579]

E. [Chapter 9] On the Bull of Florida [or the Butrol]

Thevet, volume 2, book 23, chapter 2, says that in Florida large bulls are found that the savages call *Butrol,* which have horns only

84. Figure of the Florida Bull

one foot long, having on their back a tumor or hump like a camel, their hair long above the back, of a dun color, their tail like that of a Lion. This animal is among the most ferocious known, because of which it never lets itself be tamed, unless it is stolen and ripped [when a youngster] from its mother. The savages use its skins against the cold; and its horns are very much prized, on account of the property they have against venom; and therefore the Barbarians keep some of them [i.e., horns] in order to withstand the poisons and vermin that they encounter when going through the countryside. [1579]

F. [Chapter 7] Description of the Rhinoceros

Pausanius writes that the Rhinoceros has two horns, and not [just] one alone; one is on the nose, rather large, of a black color and of the thickness and length of that of a water buffalo [or wild ox], not, however, hollow within, nor twisted, but completely solid and very heavy; the other comes out of him on top of the shoulder rather small, but very sharp. By that it is apparent that this cannot be the Unicorn, which must have only one [horn], as its name

85. Figure of the Rhinoceros

Monoceros testifies. They say that he [the Rhinoceros] resembles an elephant, and [is] almost of the same stature, except that he has shorter legs, and the nails [ungues] of cloven feet, his head like a swine, his body armed with a scaly and very tough hide, like that of the crocodile, resembling the showy trappings of a war-horse.

Festus says that some [persons] think that this is a wild Egyptian
ox. 232

[Chapter 8]

(There is one thing worthy of being noted in this animal called the Rhinoceros, [and] that is that it has a perpetual enmity with the Elephant; and when it wants to get prepared for battle, it sharpens its horn against a rock, and always tries to get the Elephant by the belly, which is a great deal more tender than his back; it [the Rhinoceros] is as long as the Elephant, but even so its legs are lower, and its hair is the color of boxwood speckled in several places. Pompey, as Pliny writes, chapter 20, book 8, had
one first shown in Rome.) [1573, 1579] 233

APPENDIX 2

From THE BOOK OF TUMORS

In the 1579 edition, the following passages constituted the final parts of *Des Monstres*. They are here reconstructed according to the footnotes of Malgaigne (vol. 3, p. 792) and texts (vol. 3, p. 792, as well as vol. 1, pp. 356–57 and 353–54). In 1585 these sections were transferred to the book *On Tumors in General.*

Histoire digne, or, Story worthy of being carefully considered both by Physicians and by Surgeons

Isabeau Rolant, the wife of Jehan Bony, living at the rue Monceaux near St. Gervais, where the sign of the Red Rose hangs, at the age of sixty years, on the xxii October 1578 was opened (having died) by the order and in the presence of monsieur Milot, Regent Doctor and Reader at the schools of Medicine; and the Pancreas and Mesentery was [*sic*] found to be of an awesome and almost unbelievable largeness, weighing ten and a half pounds, completely scirrhous on the outside; and it adhered only to the vertebrae of the Lumbar region, and, in the front to the Peritoneum, which was likewise scirrhous and similar to a cartilage; of which a dissection and demonstration were made the next morning at the home of the aforementioned Sir Milot, in the presence of Monsieur de Varades, Physician and Advisor of the King and Dean of the faculty of Medicine; monsieur Brouet, Physician of the King and [also] of Monseigneur the Cardinal of Bourbon; messieurs Cappel Marescot, Arragon, Baillou, Riolan, [all] Regent Doctors of the faculty of Medicine; Pineau, master Surgeon; ([plus] Rebours [1579]); I attended it also, and several others, and in it there were
234 found countless abscesses, each having its kystis, some full of a liquor similar to olive oil, others to honey, others to melted suet others a broth, others to the albuginous humor, others to the aqueous humor; in short, for as many abscesses as there were there was as much diverse matter in them.

Now it is to be noted that said tumor had begun eight years and more before, and had grown more and more, yet without pain; in fact, the Mesentery has no feeling; and said Rolant woman had her animal, vital, and free natural actions (or practically) as if in full health, except two months before dying, when she took to her bed because of a constant fever, which did not leave her until her death, as also because of the heaviness of her load, which she said felt as if it would become uprooted. In fact, it was found to be adhering only to the vertebrae of the lumbar region and of the Peritoneum, as I said above, and not at all to the bowels and other parts, to which it is normally attached. So that, falling upon the bladder, and pressing upon it, it caused her difficulty in urinating, as, also, pressing on the bowels, it caused her difficulty to make a stool, so that she did it only if taking some medicine by mouth. As for enemas, they could not enter; suppositories did her no good. She also had difficulty in breathing, on account of the compression of the Diaphragm. Some of the Physicians which were caring for her had the opinion that it was a mole, others that it was dropsy; 235
in fact, dropsy did ensue, and a bucket of water and more was drawn from her body. Which occurred principally because of the liver, which was completely scirrhous and filled with abscesses, both within and without. The spleen was found also to be completely rotted, the bowels and omentum liquid and spotted; in brief, there was no part [still] intact in the entire belly.

An almost identical story

Concerning this, said sieur Milot told me about having read an almost identical story, written by Jean Philippe Ingrassias, learned Physician of Sicily, [who] in his book that he composed, entitled *De tumoribus praeter naturam,* volume 1, chapter 1, book 1, tells of a certain Moor who was hanged for theft, and on whom a dissection was performed in a distinguished and large company, at which the aforementioned Ingrassias presided, and in the Mesentery were found seventy small scrofulous tumors, each one of them having its kystis, which [tumors] adhered to the external membrane of the intestines, some full of a hardened matter, similar to plaster, and others to a viscous and sticky matter, and [still] others of a more liquid matter. And it is to be noted that the other parts of the Moor's body were very healthy and whole, chiefly the liver and spleen, as the aforementioned author tells. From which he gathers that Nature, sending all the excrements of the body upon the Mesentery and neighboring parts, had purged and cleaned the others and maintained these [latter] in health; with the result that said Moor was, for as long as he lived, either not very or not at all sick.

Which is the opinion also of monsieur Fernel, book 6, chapter 7, in which he treats diseases—causes and signs—of the Mesentery and Pancreas: namely, that such abnormal abscesses and tumors are made by a discharge of nature, which, being pressed by many excrements, sends them toward the Mesentery and Pancreas, as if within a cloaca or gutter of the whole body; for those who are intemperate and excessive in eating and drinking amass a great quantity of all sorts of pituita [phlegm] and cholera [bile], which, if it does not purge itself in time and place, increases in the ventricle, liver, and spleen. Nature hard after sends it into the Mesentery and Pancreas, through the branches, which from the portal vein are inserted and lost in the Pancreas and Mesentery. Wherefore it is not without cause nor without good reason and experience (in view of the fact that these parts receive so many excrements) that said Fernel affirms and assures that he has often found the cause and the seat of choleric flux or dysenteries, hypochondriac melancholias, diarrheas, atrophies, languors, slow and erratic agues, in these parts.

To return to the subject at hand, said Ingrassias tells the above story as a confirmation of what he wrote about having read in Julius Pollux, that scrofulae are engendered sometimes in the Mesentery. Which is in conformity with the doctrine of Galen, who claims these scrofulae to be nothing other than scirrhous and hardened kernels.

Now, that there are several kernels in the Mesentery, that has been previously demonstrated in our *Anatomy.* We have seen similarly some women, having deceased, who had their womb completely scirrhous and of the thickness of a man's head, which were thought to be a *mole,* which was not the case; also one [sometimes] sees the womb to be scirrhous in one part only, all of which scirrhuses are incurable.

APPENDIX 3

From THE SICKE WOMANS PRIVATE LOOKING-GLASSE

The following passage is taken from John Sadler's *The Sicke Womans Private Looking-Glasse* (pp. 133–42). The excerpt illustrates the extent to which the treatment of monsters had become a genre. Note the similarities between Sadler and Paré in their concerns, sources, and case histories. Sadler intended that the woman afflicted by some gynecological problem or concerned about producing a "monster" might read this book in order to avoid having to ask the physician or surgeon embarrassing questions.

Chapter 13, "Of the generation of monsters"

Hernius By the Ancients monsters are ascribed to depraved conceptions; and are defined to bee excursions of nature, which are vitious one of these foure wayes. In figure, situation, magnitude, or number.

Ruffius In figure, when a man beares the character of a beast, as did the monster in *Saxonia,* which was borne about the time of *Luthers* preaching.

In magnitude, when one part doeth not equalize with another, as when one part is too bigge or too little for the other parts of the body; and this is so common amongst us, that I need not produce a testimonie for it.

Conradus; In situation, as if the eares were on the face and the eyes on
Licostenes the brest or legges, of this kinde was the monster borne at *Ravenna* in *Italy,* in the yeare 1512.

In number, when a man hath two heads or foure hands, of this kinde was the monster borne at *Zarzara* in the yeere 1540.

I proceed to the cause of their generation, which is either Divine or Naturall. The Divine cause proceeds from the permissive will of God, suffering parents to bring forth such abominations, for their filthie and corrupt affections which are let loose unto wickednesse,
Gellius like brute beasts that have no understanding. Wherefore it

was enacted amongst the ancient Romans, that those which were any wayes deformed should not be admitted into religious houses. And S. *Hierome* in his time grieved to see the deformed and lame offered up to God in religious houses. And *Kekermane* by way of inference excludeth all that are mishappen from the presbyteriall function in the Church: and that which is of more force than all, God himselfe commanded *Moses* not to receive such to offer sacrifice amongst his people; and hee renders the reason, Least hee polute my sanctuaries: because the outward deformity of the body is often a signe of the polution of the heart, as a curse layd upon the child for the parents incõtinency. Yet there are many borne depraved which ought not to bee ascribed unto the infirmity of the parents. Let us therefore search out the naturall cause of their generation, which (according to *Aristotle* and *Avicen* which have dieved into the secrets of nature) is either in the matter or in the agent, in the seed or in the wombe.

Levit.

The matter may bee in fault two wayes, by defect or by excesse. By defect when as the child hath but one legge or one arme. By excesse, when it hath three hands, or two heads.

The agent, or wombe may be in fault three wayes. First, in the formative facultive, which may be too strong, or too weake, by which is produced a depraved figure. Secondly, in the instrument or place of conception, the evill conformation or disposition whereof, will cause a monstrous birth. Thirdly in the imaginative power at the time of conception, which is of such force that it stamps the character of the thing being imagined upon the child: so that the children of an adultresse may be like unto her owne husband though begotten by another man; which is caused through the force of the imagination which the woman hath of her owne husband in the act of coition. *Aristotle* reports of a woman, who at the time of conception beholding the picture of a Blacke-more, conceived and brought an *Ætheopian*. I will not trouble you with any more humane testimonies; but I wil conclude with a stronger warrant. Wee read how *Iacob* having agreed with *Laban* to have all the spotted sheep for the keeping of his flocks; to augment his wages, tooke hasell rodds, and pilled white strakes in them and layd them before the

Cardanus

Gen.

sheep when they came to drinke, and the sheep cuppling there together, whiles they beheld the rods conceived, & brought forth spotted young.

The Imagination also workes on the child after conception: for which wee have a pregnant example of a worthy gentlewoman in *Suffolcke,* who being with child and passing by her butcher killing of meat, a drop of blood sprung on her face, whereupon she said, that her child would have some blemish on the face, and at the birth it was found marked with a red spot.

Some are of opinion, that monsters may be ingendred by some infernall spirit. Of this minde was *Egidius Facius* speaking of that deformed monster borne at Cracovie. And *Hieronimus Cardanus,* writeth of a maid, which was got with child, by a Divell, shee thinking it had been a faire young man. The like also is recorded by *Vincentius,* of the Prophet *Marlin,* that he was begotten by an evill spirit.

But what a repugnancie would it bee, both to religion and nature, if the Divells could beget men, when we are taught to believe, that not any was ever begotten without humane seed except the Sonne of God. The Divell then being a spirit having no corporall substance, but in appearance, and therefore no seed of generation; to say that hee can use the act of generation effectually, is to affirme that hee can make something of nothing, and consequently the Divell to be God, for creation solely belongs to God alone. Againe if the Divell could assume to him a dead body, and enliven the faculties of it, and make it able to generate (as some affirme hee can) yet this body must beare the image of the Divell; and it is against Gods glory to give permission so farre unto him, as out of the Image of God to rayse up his owne of-spring. In the schoole of nature wee are taught the contrary, viz. that like begets like; therefore of a Divell cannot man be borne. Yet it is not Ruffius denied, but that Divells transforming themselves into human shapes, may abuse both men and women, and with wicked people use the works of nature. Yet that any such conjunction can bring forth a human creature is contrary to nature and religion.

SUGGESTED IDENTIFICATIONS

Paul Delaunay, in his two books, *Ambroise Paré, naturaliste* and *La Zoologie au XVIe siècle,* has paved the way for discerning in *Des Monstres* some modicum of "scientific" authenticity. In these two books he has proposed identifications of certain known birth defects and of certain animals discussed by Paré. We have reproduced Delaunay's commentary from *Ambroise Paré, naturaliste,* together with corrections, additions, and amplifications by Philip D. Pallister, M.D., and William B. Jackson, D.Sc.

Fig.

2 *Symèle,* Delaunay, pp. 62–66. Symmelus with tripodia (3 feet) or monopodia with further malformation of the foot and hemimelia (half limbs or portions thereof). One should remember that a symmelus (a monster fetus with legs fused) may have none to four feet or portions thereof, that the genitalia are ambiguous, and that the upper limbs could well be hemimelic (some portions missing) with rudimentary fingers. The eye at the knee and the horn do not strike a chord of recognition. PDP.

3 Dicephalus dipus dibrachius (2 heads, 2 feet, 2 arms). PDP.

4 Spondylodymus (fused at spine). PDP.

5 *Heterotypien héteradelphe,* Delaunay, p. 64. Heterotypic heteradelphia (one twin more developed than the other), or more likely a thoracopagus epigastricus, a conjoined twin—one parasite attached to the autosite at the thoracoepigastric region. PDP.

6 *Encéphalien podencéphale,* Delaunay, pp. 62–66. Posterior encephalocele (podencephalic, i.e., hanging on a peduncle and no calvaria) or anencephaly with overgrowth of neural tissue (less likely) and with genu recurvatum. Malgaigne felt that this podencephalic monster had had "fanciful" features added. The proper translation regarding the legs clearly indicates genu recurvatum or posterior subluxation (dislocation) present at the knees ("knees like hams") which could easily result from the neural problem. I have no comment about the hands. PDP.

8 *Monomphalien ectopage* (one umbilicus), Delaunay p. 63. Ectopagus—fused at sides of bodies. PDP.

9 *Eusomphalien pygopage,* Delaunay, p. 62. Syncephalus ectopagus (single face and head). PDP.

10 *Eusomphalien métopage,* Delaunay, p. 62. Cephalopagus (metopagus—fused at the metopic area). The discussion of surgical separation of the metopagic twins in 1495 at age 10, after the death of one, is not much different from the technological feat in 1979, when similarly joined twins were separated. Both are living. The first successful separation of Siamese twins was done by Lester R. Dragstedt at the University of Chicago on 29 March 1955. PDP.

11 *Ischiopage,* Delaunay, p. 63. Ischiopagus (fused at the pelvis). Malgaigne commented that these are likely both girls and an enlarged clitoris is mistaken for a penis. I agree. Ambiguous genitalia would be most likely, and chromosomal mosaicism and other cytogenetic gymnastics are too complicated to invoke. PDP.

12 *Monomphalien sternopage* (one umbilicus), Delaunay, p. 63. Dicephalus dipygus (anakadidymus) or thoracopagus with left tetradactyly. The four fingers of the left hand could be a reduction effect due to crowding, or a developmental field effect (improper induction of the hand). PDP.

13 *Sysomiens* (fused bodies), *derodyne* type, Delaunay, p. 64. Dicephalus with ambiguous genitalia (it is unlikely that the sexes are different). PDP.

14 *Moncéphalien déradelphe,* Delaunay, p. 64. Deradelphus (twin monster with one head and neck), tetrapus (4 feet), tetrachirus (4 hands). PDP.

15 Epigastricus paracephalus (a parasitic head attached to the epigastrium). It is unlikely it took nourishment. PDP.

16 Ischiopagus. PDP.

17 *Sycéphalien synote* (ears persist horizontally below the mandible), Delaunay, p. 64. Craniopagus or cephalopagus in a pig. PDP.

20 *Ectopage,* Delaunay, p. 63.

21 Deradelphus tetrapus tetrachirus (cf. fig. 14). PDP.

22 Dicephalus dipus monobrachius (or unilateral amelia). Likely a space limitation problem. PDP.

23 Temtamy and McKusick compare this nine-year-old boy to a case of theirs with similar absence deformities (malformations) of the limbs. I have made similar findings in a fetus borne by a diabetic mother with the caudal regression syndrome (without the hand change, however), particularly ectrodactyly with septodactyly as a reduction effect. Paré's case description and depiction of the pelvis and thighs strongly suggest caudal regression. The ears in my case were abnormal. (I note that the ears here are abnormal, unlike all other ears depicted.) PDP.

24 *Omphalosite paracéphalien; type hémi-acéphale,* Delaunay, p. 66. Possibly hemicephalus (lacks brain and calvaria) or anencephalus with partial spina bifida and redundant neural tissue posteriorly. Most likely *acardius acephalus* (facial features on head). This is always one of twins. PDP.

25 *Ectromélie bithoracique,* Delaunay, pp. 64–65. Bithoracic amelia. The facility of such individuals to use the feet is widely known and was recently featured on the television program *Sixty Minutes*. PDP.

26 The girl may well have had the Brachman-DeLange syndrome. The boy might well have represented successful salesmanship by the mother. PDP.

28 The frog-faced boy is almost certainly a case of mandibulofacial dysostosis (Treacher-Collin type?), although deafness was not mentioned. It is not a case of commissural harelip as Delaunay suggests, p. 61. PDP.

29 Arthrogryposis multiplex congenita. This is an environmental disorder—often viral and induced *in utero*—that leads to clubbed hands and feet as well as to other problems. The patients often have normal intelligence. It is very common. PDP.

30 A mummified fetus dead six years with acheiria on the right. Lithopedion (a stony or calcified fetus) is described in this chapter. PDP.

37 The half-man, half-dog child could quite possibly represent the Brachman–De Lange syndrome, which is often associated with limb reductions (here of the legs) and great hairiness, especially of the back, thighs, and lower legs. (The girl "furry as a bear—deformed and hideous" depicted in fig. 26 could well have been another example of the same condition now commonly recognized.)

43 A human "monster," perhaps a sirenomelus, or else a confusion with the dugong, i.e., *Halicore tabernaculi?* Delaunay, p. 24. Now referred to as *Dugong dugong*. WBJ.

45 Probably a "hooded seal" or *Cystophora cristata,* Delaunay, p. 33.

49 A *baudroie,* or *Lophius piscatorius,* Delaunay, p. 21. The goose fishes (genus *Lophius*) are deep-water marine species. WBJ.

51 *Phoca vitulina,* Delaunay, pp. 25–26. The common names are hair seal or harbor seal. WBJ.

52 An unidentified cetacean, Delaunay, p. 31.

55 The *Pennatula rubra* or *P. spinosa* and the *Veretillum cynomorium* Delaunay, p. 32. These are simple invertebrates, not fish, and are known as sea plumes and sea pens, respectively. WBJ.

56 The *aloés* suggests the *Plesiosaurus* of the Jurassic seas, Delaunay, p. 35.

59 The Bulampech may be the *Exocoetus evolans* or the *Exocoetus mesogaster* of the Antilles, Delaunay, p. 35.

60 A large ray or skate, Delaunay, p. 13. A cow-nosed ray belonging to the genus *Rhinoptera,* probably *R. marginata* found in the Mediterranean. These rays have a facial pattern that gives them the appearance of a split lip, i.e., a double mouth. They are recognized as creatures that frequently jump from the water's surface. WBJ.

62 A Selachian, Delaunay, p. 18.

63 A cephalopod mollusk, Delaunay, p. 18.

64 An arthropod crustacean, Delaunay, p. 23.

66 A *cachalot,* or sperm whale, Delaunay, p. 38.

69 The toucan, a *Rhamphastus,* Delaunay, p. 37.

70 A bird of paradise, or *Paradisea apoda,* Delaunay, pp. 34–35.

75 Perhaps a South American *aï* (*Bradypus tridactylus*), a toothless vegetarian of the type *Bradypus,* i.e., a three-toed sloth, Delaunay, p. 36. Now referred to as *Bradypus torquatus*. WBJ.

76 "Absolutely fanciful," Delaunay, opposite p. 33, fig. 4.

80 Note that the artist has depicted the tusks protruding from the *lower* jaw (as is characteristic of boars), but in reality tusks of sea animals protrude from the *upper* jaw. WBJ.

81 A *Trichechus rosmarus,* or manatee, Delaunay, p. 29. However, as the manatee lacks tusks, this is more likely a walrus, i.e., *Odobenus rosmarus*. WBJ.

82 Perhaps the *Oryx beïsa,* Delaunay, p. 33.

84 A bison or *Bos americanus,* Delaunay, p. 37. Now referred to as *Bison bison*. WBJ.

NOTES

INTRODUCTION

1. Lyons and Petrucelli, *Illustrated History of Medicine,* p. 9.

2. This phrase is almost inevitably associated with Paré; it is one he repeated several times in his *Journeys in Diverse Places*. Paré's humility and piety are taken to be summed up in the saying, though d'Eschevannes, among others, views it as simply a motto.

3. Emile Haag and Eug. Haag, in *La France protestante,* 8. The question is discussed more fully in note 9.

4. In 1536, when the emperor Charles V went into Provence, Paré joined the campaign of Francis I as surgeon of the Maréchal de Monte-Jan. When the latter died in 1539, Paré returned to Paris. In 1543, Paré again joined the army as surgeon, going to the camp of Perpignan in the service of the vicomte de Rohan. In 1545 he reputedly saved the life of the duc de Guise at the siege of Boulogne. He is said to have removed a lance shaft imbedded in a face wound. (But see Hamby, *Ambroise Paré,* p. 51). Paré then returned to Paris, but in 1549 he returned to Boulogne, which was again being besieged. In July 1552, he went with M. le Vendôme to Picardy, where the latter took command of the army against the Spaniards. In October of this same year, the emperor Charles V sent the duc d'Albe to cross the Rhine and lay siege to Metz, defended by the duc de Guise, and in November Paré was ordered to Verdun, to report to the Maréchal de Saint André. He took with him a large supply of medicines. Paré, under royal orders, then went to Hesdin in Picardy. Though Paré was taken prisoner, he was freed by the duc de Savoie, whose ulcerous leg he had cured. In 1554 Paré was back in Paris and from 1557 to 1559 he was again sent to several military encampments. In July 1559 after Henri II died, Paré was retained by his successor Francis II, who died a year and half later. (Some rumored that Paré had slipped a poison in his ear at the behest of Catherine de Medicis; he probably died of a mastoiditis.) But Paré was again retained, during the regency of Charles IX; later, Paré followed Charles IX's armies to the sieges of Blois, Tours, Bourges, and Rouen. He was present at the attack on the English who dominated Le Havre. He then returned to Paris. From 1564 to 1566 Paré followed Charles IX and his court across France, to Nancy, Dijon, Lyon, Montpellier, Provence, and Bayonne. When Charles IX died in 1573, Henri III retained Paré and made him a *conseiller.* In May 1588 Henri III left Paris on a campaign, but by now Paré was too old to follow. In 1590 Paris, besieged by Henri IV after the battle of Ivry, was dying of famine. It was at this point that Paré called on the archbishop of Lyons, who happened to be in the city, to intervene in the interests of peace. The siege was raised on 29 August, and Paré died, probably on December 20 of that same year.

Although Paré considered his years in the Hôtel-Dieu as the most crucial to the formation of his art, these many military experiences were also very important in his formation and growth. Many discoveries came out of the campaigns at which Paré was present as army surgeon, which were, in fact, to serve him well in his Parisian practice.

5. Before Paré's discovery, gunshot wounds were routinely cauterized with boiling oil. At the siege of Turin in 1536, Paré ran out of hot oil and gave the remaining patients a dressing of egg yolks, oil of roses, and turpentine. During the hours that followed, he noted that the soldiers who had not received the conventional treatment were recovering better than those who had. After this he experimented and concluded that gunshot wounds were not in and of themselves poisonous and did not require cautery.

6. Stephen Paget, *Ambroise Paré and His Times,* p. 57. *Apologie,* Malgaigne, ed., *Oeuvres complètes* 3:687.

7. Malgaigne's Introduction to the *Oeuvres complètes* of Ambroise Paré. This long preface to Paré's *Oeuvres complètes* was translated in 1965 by W. B. Hamby (Malgaigne, *Surgery and Ambroise Paré*). See also Janis L. Pallister, "Fifteenth Century Surgery in France: Contributions to Language and Literature."

8. Paget, *Ambroise Paré,* p. 175.

9. Haag and Haag, *La France protestante* 8:130–31. None too objectively they write that Malgaigne had suggested that *perhaps* from 1562 to 1575 Paré had been a Huguenot (which may have explained why during that period someone had attempted to poison him while he was in Rouen by putting arsenic in his cabbage). The Haags, however, have no need of a "perhaps." They write, "we have not had need of this proof to couch our conviction; we only needed to glance over the writings of Paré; at every moment one feels in this honest, pious, charitable soul, full of confidence in God, the Huguenot palpitation." And, furthermore, they add, Paré's tolerance in treating the wounded of both sides proves it, as does the fact that his quotations come from the Huguenot translation of the Bible, and the fact that one of his children received the Old Testament name Isaac, and so on.

10. D'Eschevannes, *La Vie d'Ambroise Paré,* chap. 4; Paget, *Ambroise Paré,* pp. 252 ff.

11. Paget, *Ambroise Paré,* p. 192.

12. Ibid., p. 200.

13. Pierre de L'Estoile, *The Paris of Henry of Navarre,* p. 196.

14. Paget, *Ambroise Paré,* p. 231.

15. On humanist medicine in France (including the work of Dubois, Fernel, Ruel), see Arthur Tilley, *Studies in the French Renaissance,* pp. 143 ff. See also Emile Callot, *La Renaissance des sciences de la vie.*

16. In stressing moderation, Paré urged an important doctrine of the ancients that was, indeed, to become the fundamental dogma of French classicism. However, one should not forget that moderation was also a commonplace of medieval medicine.

17. Jurgis Baltrušaitis, *Réveils et prodiges,* pp. 330–31.

18. Paré, *Oeuvres complètes,* vol. 1, p. cclxxxiii.

19. These criticisms were no doubt in Pierre de Ronsard's mind when he wrote the following quatrain, which adorned Paré's *Oeuvres:*

Un lit ce liure pour apprendre,
L'autre le lit comme enuieux:
Il est aisé de le reprendre,
Mais mal-aisé de faire mieux.

(That is, one person reads this book for instruction, another reads it because envious; it is easy to criticize it, but hard to surpass it.) Ronsard also wrote a sonnet in praise of Paré, which Malgaigne quotes together with this quatrain, I, ccc.

20. Jean Céard, *La Nature et les prodiges,* p. 305.

21. Ibid., p. 306.

22. Ibid., pp. 312–13.

23. Ibid., p. 317.

24. Ibid., p. 292.

25. Other translations into English cited by Malgaigne include: *Cure of Wounds made by Gunshot,* Walter Hammond (London, 1617) and *An Explanation of the Fashion and Use of Three and Fifty Instruments of Chirurgery* (London, 1634) "Guthered out of Ambrosius Pareus, the famous French chirurgion, and done into English, for the behoofe of young Practioners in chirurgery, by H. C." The translator of this latter says in the preface: "I could not chuse a better author." This work can be found in facsimile reprint (New York: Da Capo, 1969.) Moreover, Doe (*Ambroise Paré: A Bibliography,* pp. 215, 230) mentions *Selections from the Works* (of Paré) by Dorothea Waley Singer (London, 1924). These are merely excerpts taken from the Johnson translation, as are the passages from Paré in "Epoch-Making Contributions to the Study of Syphilis" published in the *American Journal of Syphilis* (1923–24). (See Doe, p. 231.) Similar excerpts from Johnson published in various medical works are described by Doe, p. 232. In addition, Doe points out (p. 213 and also p. 225) that Paré's account of the siege of Metz was published in T. C. Minor's English translation in the *Cincinnati Lancet-Clinic* (1897). Malgaigne mentions without commentary the citation of a 1578 English edition of Albrecht von Haller in his *Bibliotheca Chirurgica* (2 vols.; Basel and Bern, 1774–75). Doe (p. 166) views this as a bibliographic ghost arising from an error (Johnson, 1678?).

26. See Erwin Panofsky, "Artist, Scientist, Genius," in *The Renaissance, Six Essays,* pp. 123–82. His idea is that the rise of the observational natural sciences was predicated upon the rise of representational techniques.

27. See J. Pallister, "Fifteenth Century Surgery." See also Petit de Juleville, *Histoire* 3:822–53.

28. See *Oeuvres complètes,* vol. 1, p. cccxxxiv. See also Petit de Juleville, *Histoire* 3:520–21.

29. Malgaigne writes, "Il rit avec sa plume à la manière de Rabelais" (*Oeuvres complètes,* vol. 1, p. cccxxxiv).

30. Callot (*La Renaissance des sciences de la vie*) points out that Paré, like Palissy, had as his method "practice and common sense" and that while these two men consulted authority, they were also disposed to question it. For Callot,

Paré is "the reformer of surgery and the greatest practitioner of his time" (pp. 23–24).

TEXT AND APPENDIXES

1. Among the moderns he followed, Paré mentions Pierre Boaistuau, author of *Histoires prodigieuses* [Marvelous stories] (1560), whose name he spells "Boistuau" throughout, and Claude Tesserant, author of another *Histoires prodigieuses* (ca. 1567). Among the ancients (read in French or Italian translation) are Hippocrates (ca. 460–ca. 370 B.C.), the father of medicine; Galen (A.D. ca. 130–ca. 200), author of 83 extant medical treatises; Empedocles (ca. 495–ca. 435 B.C.), philosopher and early physicist; Aristotle (384–322 B.C.), Peripatetic philosopher, here cited from the work *Les Problèmes d'Aristote* (1554); Pliny (A.D. ca. 23–79), author of the great *Historia naturalis* or *Natural History.* Paré often referred to Lycosthenes (Theobald Wolfhart, called Conradus), author of a work entitled *Prodigiorum ac ostentorum chronicon* (Basel, 1557). Although Paré places him here among the ancients, it is clear from chap. 38 that he cannot have seriously thought Lycosthenes was an ancient. Saint Paul's Epistles are scarcely cited in *Des Monstres,* and reference to the Apocalypse of Esdras the Prophet is restricted to one quote, in chap. 3, which Céard claims is apocryphal. Paré's use of Saint Augustine (A.D. 354–430) is largely confined to the commentary on monsters found in the *City of God.*

2. Spital beggars, often malingerers, are those who station themselves at public entrances or beg from door to door.

3. Here in the French text is an example of Paré's vacillation between the common term *fleurs* and, four lines above, the technical term *sang menstruel.* Although in Renaissance English texts one also finds the term "flowers" to refer to the menstrual flow, this translation has been rejected as obsolete. Béroalde de Verville believed the word related to the flow and not to seed (*Palais des curieux,* chap. 36, p. 276).

4. According to Malgaigne, the story is taken from Lycosthenes' *Prodigiorum,* p. 438.

5. Malgaigne cites Lycosthenes, *Prodigiorum,* p. 517, as the source, saying he in turn borrowed from Jacob Rueff, *De conceptu et generatione hominis* (Zurich, 1554). This is the famous "Ravenna monster." It is mentioned early in monster literature, and Céard, in his critical edition of *Des Monstres* (pp. 154–55), studies various illustrations of it (fig. 86).

6. The word *ventre,* usually "belly," has various meanings, including "womb," "breast," "chest," and "heart." Except where the meaning is clearly "womb," I have consistently used "belly" to translate it.

7. Caelius Rhodiginus, *Lectionem Antiquarum* [Readings from the Ancients] (Basel: Froben, 1542). Paré refers to this work as *Antiques leçons,* for he certainly read it in translation.

8. Malgaigne refers to the famous Rita Christina, studied by M. Geoffroy Saint-Hilaire in his *Histoire des anomalies de l'organisation* (Paris, 1836), 3:166.

86. Alternate Figures of the Ravenna monster, after Boaistuau (top) and Lycosthenes (bottom)

Malgaigne says that the picture was taken from Boaistuau, folio 128v°, and that Boaistuau in his turn had taken it from Lycosthenes, *Prodigiorum,* p. 565.

This monster, like the one from Ravenna, has a long history in the literature, carefully traced by Céard (*Des Monstres,* pp. 155–57). As several stories become fused, so do the geographical locations. Céard refers to the subjects of these entangled accounts as "travelling monsters."

A variant illustration (fig. 87) is also offered by Céard.

9. [If by "parents," Paré means "father and mother," and not "relatives," this part of the story is unlikely, for the mother could not have survived such a delivery. PDP]

87. The double monster of Rhodiginus and that of Lycosthenes

10. Malgaigne states that this figure is taken from Lycosthenes, *Prodigiorun* p. 490.

11. According to Malgaigne, Paré manifestly copied this monster from Boais tuau, who claimed to have seen it in Valencia in 1530 and was thus describin it from memory. Lycosthenes (p. 524) claims one very like this was seen in Savc in 1519. Malgaigne refers to a nineteenth-century monster known to him th was apparently almost identical in all respects.

12. The Piedmont is, of course, the region of Italy called Piemonte, of whic Turin is the capital. Quiers is the village of Chieri, about six miles outside Turir A league was a variable measurement, giving from one to four miles. Se Cotgrave.

13. Is the statement that the woman was "good" (*honneste*) meant to sugges that, in spite of her moral character, she was afflicted with this offspring, an hence, implicitly, that monsters may well occur where there has been no sin The adjective justifies our seeing Paré's approach to the phenomenon of birt defects as "scientific."

14. The "chaperon," or "maiden's hood," worn by the bourgeoises, document the dress of the period; here and in many other passages of Paré's oeuvre, w may look to him for information on French Renaissance culture, style, and mores For Paré as historian, see Doe, pp. 213–15.

15. Malgaigne supposes it was a question here of a posterior encephalocele and that other phenomena were added or distorted through fear or credulity Malgaigne had suppressed the illustration as being "manifestly ridiculous an imaginary" despite Paré's claim that it was "true to life."

16. Although Paré may really mean "dry," one might observe also that i Renaissance French, *sec, sèche* can refer to the skeleton. (The word skeleto

self is akin to the Greek *skleros,* "hard" or "dry.") Compare Montaigne, *anatomie sèche,* "skeleton" (*Essais,* book 1, chap. 20).

17. Sebastian Munster (1489–1552) was a German monk, scholar, and geographer, whose chief work, *Cosmographia universalis* (1544), was translated into French and added to by François Belleforest in 1575.

18. Malgaigne states that the story and the figure are from Lycosthenes, *Prodigiorum,* p. 504, and not from Sebastian Munster.

19. Malgaigne supposes this monstrosity to have been observed in reality. As often the case, Malgaigne's own note is almost as interesting as the original text. He claims here that most monsters are female and that perhaps here, as often happened, the elongated clitoris was mistaken for a penis.

20. [Fig. 14 (taken from Boaistuau) and fig. 21 are identical, except for their size and their depiction of the genitalia. PDP]

21. Malgaigne tells us that this story is taken from Lycosthenes (*Prodigiorum,* p. 521), who dates it 1516 and who probably took the story himself from Rueff, *De Conceptu,* p. 44. All these stories are probably imaginary, says Malgaigne, because no such monstrosity has ever been authenticated.

22. Malgaigne attributes the figure to Rueff (*De Conceptu,* folio 7), but there it is applied to an English monster. Malgaigne claims the same figure was copied by Lycosthenes, *Prodigiorum,* p. 619.

23. According to Malgaigne, the story is taken from Lycosthenes, *Prodigiorum,* p. 644.

24. Albucrasis, i.e., Albucasis, an Arab physician of the ninth century A.D. His *Surgery* was published in Venice in 1500, with that of Guido.

25. The faulty syntax of Paré's sentence has been preserved.

26. Cromerus, or Martin Cromer (1512–89), Polish historian, author of *De origine et rebus gestis Polonorum* (1558).

27. Here is a virtuous Polish lady, like the good, honest woman of the Piedmont, who despite her upright character, is giving birth to a monstrous litter! On this anecdote from Cromer, see L. J. Rather, "Ambroise Paré, the Countess Margaret, Multiple Births and Hydatidiform Mole." See also L'Abbé Gachat d'Artigny on multiple births, *Nouveaux mémoires* 4:34–36.

28. Francesco Pico della Mirandola (1469–1533), author of *Hymni heroici* (1511). Dorothy is a famous case, taken, as Malgaigne indicates, from Lycosthenes, who in turn quotes from Pico della Mirandola.

29. Malgaigne suppressed this picture as being "pure fantasy," while asserting that it was copied from Lycosthenes. After this figure in the editions of 1573 and 1575 came the story of the epitaph of Yolande Bailey which then was moved to chap. 44, the "Livre de la Generation" (Malgaigne, 2:736). The figure is found in a recent book, Lyons and Petrucelli, *Medicine: An Illustrated History,* p. 393.

30. Here is an excellent sample of Paré's pedagogical method. He uses the technical term and then explains it for the uneducated barber or barber-surgeon and the apprentices.

31. A curious footnote to the history of laws governing "monsters" of ancien and Renaissance times. One might add that hermaphrodites and other monster served during antiquity and the Middle Ages as a point of departure for mora observations. See, for example, Jacques de Voragine in *La Légende dorée,* wher monsters are far from lacking (1:33, 2:444). See also Pliny, *Nat. Hist.,* boo 7, chap. 6.

32. Malgaigne, disapproving, suppressed the illustration (fig. 19), while in dicating that one finds the principal features of it in Caelius Rhodiginus, *Lectionem* p. 11.

33. The Palatinate was a former political division of Bavaria, or the Rhineland ruled by a palatine. The word "burg" is taken to mean a fortified town.

34. Malgaigne judges these to be female Siamese twins, taken to be herma phrodites because of the elongated clitoris. According to Malgaigne, the stor and the figure are taken from Lycosthenes, *Prodigiorum,* p. 496.

35. Malgaigne states that this figure is faithfully copied from Rueff. Th edition of 1573 had a paragraph at this point regarding the imaginary monste which was subsequently added to chap. 3. There was a long passage in the 157 edition on the *nymphes* (folds adjacent to the meatus urinarius on the female) the passage was lengthened in 1575, shortened in 1579, and finally moved i 1585 to the end of chap. 34, book 1 of *De l'Anatomie* (Malgaigne, 1:168–69) A reconstruction of this passage follows:

> Moreover, at the beginning of the neck of the womb is the entrance and crack of the woman's "nature," which the Latins call *Pecten;* and the edges, which are covered with hair, are called in Greek Pterigomata as if we were to say wings, or lips of the woman's crown, and between these are two excrescences of muscular flesh, one on each side, which cover the issue of the urine conduit; and they close up after the woman has pissed. The Greeks call them *nimphes,* which hang and, even in some women, fall outside the neck of their womb; and they lengthen and shorten as does the comb of a turkey, principally when they desire coitus; and, when their husbands want to approach them, they grow erect like the male rod, so much so that they can disport themselves with them, with other women. Thus they render them very shameful and deformed being seen naked, and with such women one must tie them and cut what is superfluous because they can abuse them, the surgeon taking care not to incise too deeply, for fear of a great flow of blood, or to cut the bladder, for afterward they couldn't hold their urine, but it would flow out drop by drop (a very monstrous thing that is done to the *nimphes* of some women).
>
> Now that these women, who by means of these caruncles or *nimphes,* abuse one another is a thing as true as it is monstrous and difficult to believe; [it is] confirmed, nonetheless, by a

memorable tale drawn from the *History of Africa,* by Leo Africanus [Arab geographer, 1483–1532?]. Among the prophets who are in Fez, the principal city of Mauretania in Africa, there are certain women who, giving the people to understand that they have familiarity with demons, perfume themselves with various scents, pretend that the spirit has entered their body, and through a change in the voice give [one] to understand that it is the spirit that is speaking through their throat. Then in great reverence the people leave a gift for the demon with them. African learned-men call such women *Sahacat,* which in Latin gives *Fricatrices* because they rub one another for pleasure, and in truth they are afflicted of that wicked vice of using one another carnally. Wherefore if some very beautiful woman goes to interrogate them, for a payment in the name of the spirit they ask her for carnal copulations. Now some of them are found who (having taken a liking to this sport and attracted by the sweet pleasure they take from it) pretend to be sick and send for these prophetesses, and most frequently send the message by their very husbands: but in order to cover up their wickedness better, they give the husband to believe that a spirit has entered into his wife's body, whose health being at stake, he must give her permission to go among the prophetesses; whereupon, the good and unsuspecting husband consenting to it, prepares a sumptuous feast for this whole venerable band, at the end of which they dance, then the wife has leave to go where she deems proper. But there exist a few husbands who, cleverly perceiving this ruse, get the spirit driven out of their wives' bodies with a good hard clubbing. Also, others—giving the prophetesses to understand that they are detained [or impeded] by the spirits—deceive them by the same means as the prophetesses have done [with] their wives. That's what Leo Africanus writes about it, assuring us in another place that in Africa, there are people who go through the city like our castrators [or, gelders, spayers] and make a trade of cutting off such caruncles, as we have shown elsewhere under *Surgical operations.*

36. Amatus Lusitanus was a Portuguese physician (1511–68), author of the *urationum medicinalium Centuria secunda* (Venice, 1552). Malgaigne observes at this is indeed the thirty-ninth story of the *Centurie deuxième* (*Centuria secunda*) f Amatus Lusitanus.

37. This passage was written in 1573, during the reign of King Charles IX.

38. This voyage was with King Charles, in November, 1573. Compare the tory of "Marie Germain" with that of Montaigne in his *Journal de Voyage* (where, s Rat observes, Montaigne mentions Paré) and also in his *Essais,* I, 21 (Rat,

1:101, 694, n. 237). Other comparisons between Montaigne and Paré occur this same essay, especially on p. 109. See also Rat's notes 258 and 261, p. 69

39. The conclusions Paré reaches are characteristic of much medieval a Renaissance thinking, which formulated perfection in terms of the male valu and attributes, and imperfection in terms of the female. Of course, ideally, th are fused together in the platonic androgyne. In the Club Français du Liv edition of Paré's *Oeuvres,* which includes *Des Monstres,* we find the followi observation: *L'homme est successeur d'Adam, créé à l'image de Dieu, tandis que femme n'a, comparée au mâle, que valeur et fonction partitives ou complémentair* "Man is the successor of Adam, created in the image of God, whereas woma when compared to the male, has only a partitive or complementary value ar function" (p. 365).

40. Malgaigne indicates that a similar figure is found in Rueff, *De Concep* folio 49 v°), and in Lycosthenes, who apparently copied it from Rueff. Th figure, often placed at the end of this chapter, in fact supplements commenta found here.

41. Identified by Céard (p. 169) as the present-day Villefranche-du-Queyra in the department of Lot-et-Garonne.

42. Malgaigne writes, "This figure, with the text which belongs with it, w added in 1575, and the author had put in the margin this naive exclamatio 'A very monstrous thing, to see a woman without a head.' One might note th the French text seems to say first of all that the monster itself was given to Pa by Hautin, but that later it is obvious that it was only the figure." Malgaig had suppressed the posterior view. In the Johnson translation Hautin has clear given to Paré the *figure* which he had got from Fontanus, a physician "Angolestre."

43. According to Malgaigne, Rueff gives exactly the same figure, but witho the tools (*De Conceptu,* folio 43). Lycosthenes copied Rueff's figure, adding t whip, hatchet, dice (*Prodigiorum,* p. 536). Compare Montaigne's descriptions armless persons (I, XIII), vol. 1, p. 115 (and p. 696, n. 272).

44. [The child described may be a case of osteogenesis imperfecta or tha atophoric dwarfism. PDP]

45. [Although we cannot accept the explanation Paré gives for the whi daughter, it is worth pointing out during this discussion of albinism that a hi mutation rate occurs in African blacks. Three paragraphs later Paré returns albinism occurring in rabbits and peacocks confined in a pen with a whi background. Today, of course, we would attribute this to a confined breedi population of animals which carry the recessive gene. PDP]

46. Malgaigne states that these last two stories were borrowed directly fro Boaistuau (even including the attribution to Damascene and the two figur which Malgaigne suppressed), folio 14. Montaigne also speaks of this monst (*Essais,* I, 21).

47. According to Malgaigne the story attributed to Hippocrates is probab apocryphal. Yet Céard identifies it as coming from *Opera,* section III, "De natu pueri," perhaps through Sylvius.

48. The word *more,* or *maure,* in Renaissance French designated a black-a-oor, or dark-skinned person, specifically an African negro. The story Paré tells ere, taken by Boaistuau "from Hippocrates," is, according to Malgaigne (3:24), ocryphal. Montaigne briefly alludes to this same story (I, 21). According to laurice Rat, in his edition of the *Essais* (1:695), the story comes from Saint rome, as cited by various compilers of the Renaissance, and especially by Pedro lexía (Spanish historian and humanist, 1499–1551) in his *Silva de varia leccion,* 540.

49. Stecquer or Stecher is not a place, but a person's name in Lycosthenes' atin original. (Céard, ed. of *Des Monstres,* p. 165).

50. Lycosthenes, *Prodigiorum,* pp. 528–30, borrowed from Rueff, *De Conceptu,* lio 48 v°. It is interesting to note that Paré does not explain here what the other's fixation might have been. This monster was cited in anti-Catholic erature of the time, especially by Luther.

51. This figure is also missing in Malgaigne, who speculated that it was a estion of an anencephalic and noted that Paré had copied the figure from a acard, as Paré himself stated in chap. 21. Delaunay (*Ambroise Paré, naturaliste,* 61) views it as a harelip.

52. A well-known method of getting the pear to grow inside a bottle as a step the preparation of pear liqueur.

53. Diphilus might be any of several Greek physicians of that name. He is ted by Boaistuau, from whom Paré takes the story.

54. A similar figure is found in Rueff (*De Conceptu,* folio 45 v°). Malgaigne oints out that it is a question here of *pieds bots* (clubfeet) and *mains botes* lubhands).

55. For Hertages [Hertoghe], Langius, and "Bassarus" [Gassarus], see Céard, d. of *Des Monstres,* p. 167, n. 72.

56. [This may be an example of hereditary skeletal dysplasia, and the case f limp mimicry in the next sentence is perhaps an instance of familial hip ysplasia. PDP]

57. It is from such passages as this that one might conclude that the versatile aré was, among many other things, an early "speech pathologist." Obviously, eech defects (a departure from the norm) would fit into Paré's concept of the monstrous." That Paré should see stuttering as transmitted by heredity is of tmost interest.

58. Lapidaires, i.e., persons who have stones, for example, kidney stones, allstones (Huguet, *Dictionnaire,* s.v.).

59. According to Malgaigne the phrase in parentheses is a posthumous ddition.

60. [This is perhaps Paré's greatest insight. PDP]

61. I.e., *Os des Isles;* according to Cotgrave, the *os des isles* "is [the bone] yned to the transverse processes of the sacred bone; and divided into three arts; the first whereof (being the highest and broadest) retains the name, 'other two are called otherwise." (The sacrum.)

62. In French, the obsolete word *laringau* (*laryngo* from the Greek *larun* *larungos:* Huguet). Céard gives "noeud de la gorge."

63. Although Paré frequently uses the vulgar verb *pisser,* it is to be noted th here, as occasionally elsewhere, he uses the technical phrase *en urinant.*

64. *Apostème,* or "apostema," is the word usually used by Paré for absces It is defined in Cotgrave as an "inward swelling full of corrupt matter." *Apostèn* has been translated by "apostema," and *abcès* by "abscess."

65. The Collo family had several illustrious surgeons who were friends Paré's.

66. The *Treatise on Stones* is now located in the treatise called *Des Opération* As all editors point out, Tire-vit is also discussed by Paré in his "Introductic ou entrée pour parvenir à la vraye cognoissance de la chirgurie."

67. Malgaigne (n. 1, 3:31, *Des Monstres*) discusses the implication of Paré variations on this story found in Paré's own Introduction (Malgaigne, 1:28).

68. Malgaigne asserts that this is the first known case of a foreign body th had developed in the knee having been successfully extracted through incisio [However, it was more probably a large joint mouse or osteochondritis dessican PDP]

69. The phrase indicates the hierarchy or chain of command existing betwee the physician and the surgeon, which was backed by law.

70. The following version was found after the story about Dalechamps in th 1573 and 1575 editions: "I was one day called along with Monsieur Le Gran [Magnus], Regent Doctor on the faculty of Medicine and ordinary Physicia of the King, a learned man, and very experienced, to insert an anal speculu into a lady of honor who was tormented by extreme pains in the belly and the seat, and yet with no visible appearance of something's being wrong; whic was the reason why he ordered certain potions and enemas for her, with or of which she voided a stone as large as a tennis ball; and suddenly her pai were ended and she got well."

71. [Probably a parotid duct stone. PDP]

72. The medieval and Renaissance theories of the microcosm and the ma rocosm which in the 1573 edition framed this chapter, entitled "On Worms had special applications for medicine, as is reflected in Paré's remarks. See this respect Cosman and Chandler, eds., *Machaut's World,* pp. 27 and 35, n. 3 Céard, *La Nature et les prodiges,* pp. 302–3; Brabant, *Médecins,* pp. 153–6 Paracelsus, in particular, had the celebrated theory of the *mysterium magnum* (the macrocosm) versus the *cagnastrum* (or microcosm).

Early on in his *Surgery* Paré explores this question. As man is the microcosn made in the image of the macrocosm (and of God), he is called upon to domina the animals, having his title as procurer from God. He is, therefore, superi to the "brute" animals and is certainly not one himself. To think he were wou be to think like an atheist.

73. Like the preceding story of the scorpion, this one dates from 1573 b in both cases the figures (not shown here) were added in the 1579 editio

(Therefore they could not have been drawn from direct observation but, rather, from memory.)

74. Malgaigne points out that Paré has only said "a certain matter *similar to an animal*," and that a bloody stone might have presented a form approximating this description and then have been exaggerated by the artist.

75. I.e., "to eat frugally rather than well" (Huguet, *Dictionnaire*). Here, as in other writings, Paré stresses the importance of a good diet.

76. Laurent Joubert (1529–83), French physician of Montpellier, author of *Popular Errors concerning Medicine and Health Regimens,* parts 1 and 2 (1587). In 1558 Joubert also published his translation of Rondelet, whom Paré refers to frequently in his section on marine animals.

77. Malgaigne says the figure was copied from Lycosthenes, *Prodigiorum,* p. 503.

78. This last sentence appeared only in the editions of 1573 and 1575 and was restored by Malgaigne from them. Yet, it would seem that Paré realized that this sentence spoiled his first-person narrative and that he removed it for stylistic reasons.

79. This figure was omitted by Malgaigne, who deemed it too "absurd" to reproduce.

80. *Menusier,* or *menuisier:* "ouvrier qui fait des travaux menus, fins" (Huguet, *Dictionnaire*).

81. Malgaigne refers us to a similar passage in Paré's book on tumors in general (Malgaigne, 1:324).

82. This treatise was subsequently divided into two parts: "On the Plague," and "Concerning Smallpox and Leprosy." Paré's reference is to the latter work.

83. Paré wrote "two years" while Benivenius (Beniveni) had written "ten." The accomplishments of Beniveni (1440?–1502) are discussed at length by Malgaigne in his Introduction to Paré's *Oeuvres complètes,* vol. 1, pp. cxii ff.

84. Chap. 17 in the early editions begins first with the story about Monsieur Sarret, now found in chap. 52 of Paré's book *On Surgical Operations* (Malgaigne ed., 2:500) and then with the one about the Count of Mansfelt, now found in chap. 14 of his book *On Handgun Wounds* (Malgaigne ed., 2:168). These are followed by the story of Monsieur de la Croix and other observations on how pus might travel (now found in the Malgaigne ed. 2:500–502). Here is a reconstruction of the missing passages:

> Monsieur Sarret, Secretary of the King and of monseigneur d'Anjou, brother of the King, was wounded from a pistol shot in the right arm, and he underwent many side effects, such as fever, abscess, ulcers, from which a great quantity of sanies [pus] issued, and then for several days very little came out, and then he ejected it first through the seat and then through his urine, and when these ulcers [had] let off a lot, one could see no evidence of them beneath; nevertheless he is still living. Which I saw similarly happen to monsieur the Count of Mans-

felt, from his pistol wound, which he got in the left arm, the day of the battle of Montcontour. Moreover Germain Cheval and François Race (men accomplished and excellent in their art, Sworn Surgeons in Paris) and I dressed a gentleman named monsieur de la Croix (as I have written in my "Treatise on the Suppression of Urine"), who was wounded from a sword thrust into his left arm and the elbow joint, to whom a similar thing happened, and yet he died; and this because according to some people it was impossible that the sanies should make such a long circuit, to reenter from the axillary vein to the ascendant vena cava, passing near the heart, which was not infected, and from there through the interior of the liver, and from this to the vena porta, then to the mesaraic [i.e., mesentric] veins, then to the emulgents [renal artery or vein] and from these across the kidneys, then to the ureter (pores), and from these to the bladder and from said mesaraic veins to the intestines and from these to the seat. Nevertheless since we see in inanimate things such things happening, as is shown to us by the experiment of two glass vessels—called montevins—the upper one of which being filled with water and the lower with wine, when placed one on top of the other, one manifestly sees the wine mount to the top across the water, and the water descend across the wine, without their becoming mixed together, although this occurs in the same narrow conduit; all the stronger reason why we should believe this to happen in Nature, which is very provident in expelling what is contrary to her; which is manifestly shown to us by women who have recently given birth and who through their womb eject milk without [its] being mixed at all with blood, and [yet] it is necessary that it pass through the mammary veins and arteries, although they be rather small, through the connection that they have together in the middle of the longitudinal muscles of the epigaster [belly] with those of the womb. Moreover, everyone knows that the liver draws white chyle from the stomach and from the intestines through the mesaraic veins to be made into blood in this, and by these same veins [to be] sent from the blood to said intestines and stomach for their nourishment; and yet they are two opposite movements. Besides, seed is made of pure blood, and the best of these that may be [found] in the body [is] sent from all the parts to be ejected through the rod; for reproduction passes through the ejaculatory spermatic vessels, which seemingly are always full of blood; nevertheless the seed flows across [or, through it] without getting mixed in at all. Wherefore one

> must conclude with Galen that the slime made in the internal and upper parts, and far from the kidneys and from the bladder, can be evacuated by way of the urinary tract.

85. Heath translates this as "lawn grass"; Cotgrave as "quitch-grass," "couch-grass" or "dog-grass."

86. Cotgrave, "Pleuretique, as pleúre: a thinne and smooth skinne wherewith th'inside of the ribs is covered," that is, of the pleura.

87. The 1573 edition gave *Monsieur* Cabrolle. Did Paré suppress the word *monsieur* in the 1579 edition because, as Malgaigne suggests, he decided this was too great a title for a mere surgeon?

88. The Hôtel-Dieu was the hospital in which Paré received his early training as a surgeon. He always considered the experience he gained there as the most valuable to his practice.

89. Malgaigne states that it was not Valesco de Taranto (1382–1418) who wrote this work, but rather the Flemish physician Rembert Dodoens, who published some excerpts from Valesco's *Philonium* at the end of his *Observations* (1581). The Céard edition of *Des Monstres* (p. 169) attributes Paré's tale to Alessando Benedetti (d. 1525), or Benedict, who is also represented at the end of Dodoens's work.

90. The *francs-archers* made up the first regular infantry troop of France, established by Charles VII in 1448. These archers were called *francs* (or "free") because they were exempt from taxes.

91. [Paré is describing here the successful removal of a pilomatrixoma, or ingrown hair tumor. PDP]

92. "Baptiste Leon" is cited by Boaistuau, but no Baptiste Leon, author of medical treatises or commentator on miracles, is known. See Céard's edition of *Des Monstres* (p. 171) for details.

93. *Des Monstres* sometimes supplies facts concerning Paré's life, as here where we learn that he owned property near Meudon. Such details substantiate our impression that he was a man of means.

94. Paré uses the form *athéiste,* rather than *athée*. The derivation of monsters from moral deviants reaches far back into the Middle Ages and antiquity. See, for example, Jacques de Voragine in his account of Saint Pélage (*Légende dorée,* 2:443), where some of the monsters described (such as the swine with the man's face, the chicken with four feet, the man and dog joined together) appear to have been handed down to Paré and his contemporaries without much modification. Paré, however, does not usually insist on moral "ends" or "reasons" for the appearance of the monsters; he is more interested in the " 'scientific' causes."

95. I.e., Volaterranus, author of *Commentariorum urbanorum* (1506), cited by Boaistuau; Cardano, *De rerum varietate,* 1581.

Malgaigne omits this figure, saying it was copied from Lycosthenes (op. cit., pp. 502 and 656). On the confusions in this passage see the Céard edition (note 111, p. 171).

96. Cybare, or Sibaris, or Sybaris, ancient city on the southeast coast of Italy on the Gulf of Taranto; a part of Magna Graecia (750–650 B.C.). See H. H. Scullard and A. van der Heyden, *Shorter Atlas of the Classical World,* p. 9.

97. Malgaigne omitted this figure from his text. It is said to have been copied from Lycosthenes, pp. 124, 136, 371, 374, etc.

98. The figure, of uncertain origin, was deemed "absurd" by Malgaigne and therefore omitted.

99. This sentence is an excellent example of Paré's sometimes incoherent syntax.

100. It is interesting to note that Paré ultimately eliminated the words "for posterity." This figure (41) was also suppressed by Malgaigne.

101. Salt surveyor, or *mesureur de sel.* During the Middle Ages and the Renaissance most French citizens were subject to the *gabelle,* a very unpopular tax on salt, then considered a precious commodity, whose distribution was carefully regulated by the government through the salt surveyor. The tax was repealed in 1584.

102. *Privilège:* publication right or permission granted by the government. The passage shows the human being's natural curiosity about monsters, which if not "signs" (*monere* + *-strum*) are at least something to "show," as another possible Latin derivation of the word monster (*monstrare*) suggests. See Montaigne, *Essais,* III, 30 and Wolff, *La Science des monstres,* p. 13.

103. Jeremiah 10:2: "Thus saith the Lord, Learn not the way of the heathen, and be not dismayed at the signs of heaven: for the heathen are dismayed at them."

According to Paré's statements in this paragraph, it follows that he repudiates the theories regarding *Homo signorum* (or government of parts of the body by signs of the zodiac), even though they were prevalent in the Middle Ages, and were upheld by many Renaissance physicians. See the figure of *Homo signorum* from Hieronymus von Braunschweig's *Experyence of the Warke of Surgeri* [London, 1525]. Alchemy and astrology in medieval and Renaissance medicine are the object of Brabant's study, *Médecins,* part 2, chap. 2.

104. Paré's sentence contains the essence of Paul's first epistle to the Ephesians (chap. 1).

105. A merging of atheists and sodomites seems peculiar, but the phenomenon is perhaps elucidated by Le Duchat (Part II, Vol II, p. 392), who declares that the expression *albigeois sodomites* is to be explained as follows: "Bulgarians, or Albigenses, were burned as sodomites under the pretext that in France *bougre* meant both a man of the sect of the Bulgarians and a man guilty of sodomy."

106. This chapter, along with several others following that deal with the question of spital beggars with sham illnesses and diseases, serves as an excellent example of Paré as a *raconteur* and as a moralist. The *conte,* an important new genre in the Renaissance, has been widely studied in recent times; and it is clear from the rapidity and wit with which these episodes are narrated that Paré had a true talent for storytelling, as well as a love for it. As several critics point out,

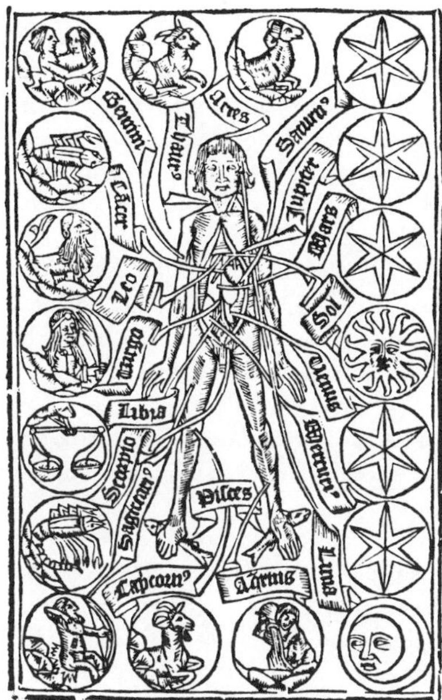

88. Homo signorum

Paré in many spots clearly imitates Noël du Fail's *Propos rustiques*. See Céard, ed. of *Des Monstres,* pp. xxx–xxxi, and also du Fail, pp. 55–61 and 172.

The unsympathetic portrayal of beggars and malingerers found in Paré calls to mind the series of 25 sketches on the same subject done a generation later by Jacques Callot, who, like Paré, saw these persons as parasites and as a serious threat to the social order.

It is, in fact, not surprising that Paré should deal with this subject, not only because of the moral "monstrosities" involved, but also as a phenomenon the physician and surgeon should truly be on guard against. Revealing in this context is Paget's statement (*Ambroise Paré and His Times,* p. 17) that in Paré's time Paris, out of a population of 150,000, counted 6000 to 7000 criminals and 8000 to 9000 paupers. (Pierre L'Estoile's memoirs flesh out the picture of Paré's Paris.) Given the small population (though estimates vary greatly), it *is* surprising that Paré observed as many abnormalities as he did. Yet everyone knew of his interest in the subject and many, even from other parts of France, contributed to the collection of curiosities he kept in his study.

107. Some scholars have used the presence of the word "temple" as evidence that Paré was secretly Protestant. But d'Eschavannes (*La Vie d'Ambroise Paré,* p. 143) points out that the word "temple" was used in Paré's time to designate a Catholic church as well as a Protestant one.

108. An astringent medicinal red clay.

109. From medieval times on into the seventeenth century, lepers were legally required to carry noisemakers to alert others of their presence. For illustrations and discussion of this see Cosman and Chandler, *Machaut's World,* pp. 6–10.

110. "Univoque et équivoque: (Terme de Médecine ancienne.) Ne s'appliquant qu'à une seule maladie" (Huguet, *Dictionnaire*). Paré clarifies this term in another work (*Oeuvres,* 22:10): "Des signes susdits les uns sont univoques, c'est à dire qui desmontrent veritablement la lepre: les autres sont equivoques ou communs." Univocal here means unmistakable, or characteristic of a particular disease.

111. The *Mal Saint Jean* was epilepsy, or chorea (Saint Vitus' Dance). The *Mal Saint Fiacre* referred to tumors or, sometimes, hemorrhoids. The *Mal Saint Main* or the *Malum Sancti Manis* was scabies, or, sometimes, leprosy. Paré usually meant scabies by this last term.

112. Were it not for Céard's assertion to the contrary, this might be taken as an example of Paré's irony. The Club Français edition of *Des Monstres* (p. 366) gives a very involved explanation of the passage, which confirms Céard's interpretation: "Les mots 'bons compagnons' ne doivent pas nous tromper car, s'ils sont 'bons', c'est entre eux, dans leur société ou 'compagnie' que les gueux le sont. Le paragraphe suivant confirme cette interprétation." (They are "good" to one another through a network.)

113. I.e., the escutcheon; the "badge" of her profession, prostitution. This may serve as an example of Paré's sometimes rabelaisian sense of humor. (Hamby, p. 100: "her ensign was displayed.")

114. *Épigastre:* "all the outward part of the bellie from the bulk to the privities." (Cotgrave.)

115. The many debts this chapter has to Pierre Martyr (via Lavater) and to Ronsard (1524–1585) are acknowledged by Paré himself and are fully studied by Céard in his edition of *Des Monstres.*

The use Paré makes of Ronsard's *Daimons* here and elsewhere indicates an admiration for the poet, who responded in kind, having written a celebrated sonnet and another piece about Paré's work and about his sufferings at the hands of jealous rivals. These poems embellished the early editions of Paré's work; they are both reproduced by Malgaigne (vol. 1, p. ccc). For the quatrain, see Introduction, note 19.

Passages of the second story of Jacques Yver's highly successful *Printemps* (1572) are not unlike Paré's developments on demons.

116. The story of Ochosias is found in 2 Kings, 1.

117. Jean Bodin (1530?–96) is an important source for Paré. The reference here is not (as Paré claims) to Bodin's famous *République* (1576), but, rather, to his *Démonomanie des sorciers.*

118. Are Paré's references accurate? (See Céard edition of *Des Monstres,* p. 177.) Saint John should perhaps not be included; Saint Luke should.

119. "Dung-hill cocks." In French, *huppes,* which Cotgrave translates "dung-hill cocks." In Johnson we read "lapwings." For the "dung-hill cock" see also

Pliny's *Nat. Hist.,* 10, 21. This is the hoopoe, or upupa, said to make its nest with human excrement.

120. The word *coquemares* offers difficulties in translation, and is here rendered by "nightriders," which might be termed a neologism. *Coquemares* are incubi; specifically those astride a mare, whence the word "nightmare." An allusion to covering or riding the female is implicit in the French term, hopefully captured in the English rendition.

121. By John Wier, the source of much material in Paré's chapters on demons. The French translation appeared in 1567: *Cinq livres de l'imposture et tromperie des diables.*

122. Paul Grillant, or Paolo Grillando, was a sixteenth-century theologian and jurisconsult of Naples and author of *Tractatus de hereticis et sortilegis.*

123. I.e., Doctors of the Church who have been canonized; e.g., Saint Augustine, Saint Bernard, Saint Thomas Aquinas.

124. Jezebel's story is found in 2 Kings 9.

125. Louis (or Ludwig) Lavater, a Protestant minister, was the author of *Trois livres des apparitions des esprits,* translated into French in 1571. The developments of this chapter are found in the mid-eighteenth-century memoirs of the Abbé Gachat d'Artigny (*Nouveaux memoires,* 4:1–4). A noteworthy parallel is to be found in Georgius Agricola's *De re metallicâ,* which contains a treatise entitled "De animantibus subterraneis." The work was published in Basel in 1564.

126. Pierre de la Pallude was a fourteenth-century patriarch of Jerusalem who wrote theological commentaries, and Martin d'Arles, a sixteenth-century theologian based in Pamplona, was the author of *Treatise on Superstition.*

127. Céard (*Des Monstres,* p. 181) claims that the theory attributed here to Saint Albertus Magnus, doctor of the Church and scholastic philosopher (1193–1280), is not to be found in his writings, but rather in the writing of his sixteenth-century editor-commentator.

128. For, as in Johnson: "How can they who neither eat nor drink be said to swell with seed?"

129. "Boulogne la Grasse en Italie." Paré is distinguishing between the French Boulogne and the Italian Boulogne (Bologna).

130. A cubit equaled a foot and a half.

131. [This paragraph details cases of pica (ingestion of foreign bodies) and of bezoar (casts of the stomach), and, in the instance of the suicide, of a trichobezoar. The persons described were no doubt mentally ill or retarded. PDP]

132. I.e., Exodus 22.

133. The story is from John Wier, *Cinq livres de l'imposture.* See also Montaigne *Essais,* III, 9. Both Wier and Montaigne take it from Saxon the Grammarian's *Canorum regum heroumque historiae,* book 14. (See Garnier edition of Montaigne 2:417, and note to this passage, 2:666.)

134. An additional paragraph was put at the end of this chapter in the 1573 and 1575 editions. In 1579 it was moved to chap. 28, where it is now.

135. This chapter (chap. 32) was lacking in the editions of 1573 and 1575.

136. "Nouer l'aiguillette: empêcher, par un sortilège, la consummation du mariage" (Huguet, *Dictionnaire*). See also note 153 below for a more elaborate discussion of this "phenomenon."

137. Pliny on Nero and magic: *Natural History*, 30, 2.

138. The Pythoness is found in Acts 16:16–18; the woman-ventriloquist, in 1 Samuel 28; King Nebuchadnezzar, in Daniel 4–5; the sorcerers of pharaoh in Exodus 7–8; Simon the magician in Acts 8:9–24.

139. Pliny on Demarchus, or Demaenetus: *Natural History*, 8, 22.

140. Essence of succinum; see Pliny *Natural History*, 37, 11–12.

141. "Not a bone of his shall you fracture" (John, 19:36). A radically different translation is given in the Club Français du Livre edition of *Des Monstres*, p. 366: "Que ne soit point broyée et privée d'usage cette bouche."

142. *From His side issued blood and water.* (John 19:33–34). [It is interesting to note that hypnotism is used today to stop nosebleeds and threatened abortion. PDP]

143. "May this fever be as easy for you to bear as the birthing of Christ was for the Virgin Mary."

144. "I exalt you God, my King" (Psalm 145:1).

145. "Teeth of the comb and tooth-shaped signs, heal me of my toothache."

146. The Johnson translation refers this passage to Pliny, *Natural History*, 28, 1. Remedies for toothache, and the like, are also found in *Natural History*, 28, 49, as well as in 32, 26.

147. A quartaine or quartan fever is one recurring every four days.

148. In French, "fait la medecine à la rate," may involve a pun. However, Johnson translates this simply, "he cures or makes a medicine for the milt." See Pliny, *Natural History*, 3, 9.

149. Johnson translates this "a leaf of Lathyris, which is a kind of Spurge." See Pliny, *Natural History*, 27, 71.

150. In French, *rapsodier*. Huguet (*Dictionnaire*) cites this very sentence from Paré and defines *rapsodier* as *coudre ensemble*.

151. The Empiricals, or Empiricists, in ancient times constituted the school of medicine that confined itself to observation and to accepted remedies that had been discovered to work, while it remained skeptical of theoretical explanations. After the sixteenth century, "empirical" came to be contrasted with "scientific" and finally degenerated into the equivalent of quack. In this context, Paré is, of course, speaking of the Empirics, or practitioners without formal training.

152. I.e., a hen ridden by a cock; related to the French word *cauchemar*, or "nightmare." The French verb *chaucher* or *cocher* ("to ride, or cover") is often seen in French Renaissance literature, e.g., Béroalde de Verville's *Le Moyen de Parvenir*.

153. In French, *nouer l'aiguillete*. Found in *Des Monstres* twice. Accounts of "knotting the codpiece" are widespread in the Renaissance. They are found in Guillaume Bouchet, *Cinquième serée,* 1, 183; E. Pasquier, *Recherches,* 5, 8; 6, 36; Le Loyer, *Histoire des spectres,* 2, 8; Brantôme, *Des Dames,* part 1: *Jehanne de France,* 8, 92. Montaigne mentions these *plaisantes liaisons* (I, 21) and Rat (1:694) cites additional references to the phenomenon in Tabourot's *Bigarrures* and Bodin's *Demonomanie*.

Esguillette nouée: A spell cast on the groom's fly (or penis). Cotgrave writes, "The charming of a man's codpeece-point, so as he shall not be able to use his owne wife, or woman (though he may use any other); Hence; *avoir l'esguillette nouée* signifies, to want erection. (This impotencie is supposed to come by the force of certain words uttered by the Charmer, while he ties a knot on the parties codpeece-point.)"

An intriguing observation is made by the editor of the Club Français du Livre edition of *Des Monstres* (p. 366) regarding Paré's apparent attitude toward the effect of this spell, considered by him to be "monstrous" and therefore unbreakable, or vice versa. Paré's treatment of this charm indicates that he recognized the domaine of magico-religious or psychic consciousness and knowledge which escaped the ordinary field of consciousness. The exact statement of this editor follows:

> Un peu plus haut, Paré se moque des incantations administrées en place de remèdes. A cette réserve près que si la médecine aide la nature à guérir le malade, elle ne vaut rien là où l'affection tient sa cause de la sorcellerie. Ainsi, «l'esquillette nouée» est-elle exactement un effet monstrueux et ne peut-on la défaire. Ceci prouve péremptoirement qu'aux yeux de Paré, un certain domaine de connaissance et d'action (que nous appellerions aujourd'hui magico-religieux) échappe *de jure* au champ ordinaire des consciences.

154. Although Malgaigne treats this section as a separate chapter, he notes that in 1573 it was merely an appendix to the chapter on incubi and succubi and in 1585 it was placed after the chapter on "point-knotters."

Although Malgaigne subsumes all subsequent chapters of the work under an appendix, we are numbering the chapters as in other editions. All figures, except that of the ostrich skeleton which Paré himself prepared, were suppressed by Malgaigne.

155. Pliny, *Natural History,* 9, 5, on tritons (mermen) and sirens (mermaids).

156. Macerie is not a place name, but became one because of Gesner's incorrect understanding of the Latin original, where the word *macerie* occurred. See Céard edition of *Des Monstres,* p. 188, n. 258. The figure, like several in this section of *Des Monstres,* is from Gesner. A few come from Thevet and Rondelet. See Céard edition.

157. Philippe Forestus (Jacopo-Filippo Foresti, or Philippe de Bergame) wa an Italian monk and historian (1434–1520), author of *Supplementum chronicoru orbis, ab initio mundi ad annum 1485* (Brescia, 1485; Venice, 1500).

Céard's commentary on this passage points up several problems (Céard editio of *Des Monstres,* p. 188, n. 259). In addition, Malgaigne tells us that the earl editions said "Martin" (Martin IV [1280–85]), and not "Marcel." No Marti was a successor of Paul III (r. 1534–49); moreover, Julius III (r. 1549–55 reigned between Paul III and Marcel II (r. 1555). (Paul IV reigned fror 1555–59.) The mistake was observed and "corrected" by some editor. Note however, that Forestus himself died in 1520, so Paré's stated source is als incorrect.

158. One must assume that the painter saw the painting in Antwerp but tha the monster was caught in the Illyrian Sea, that is, the Adriatic, *au naturel,* o *sicubi extat,* "exactly as it was." The fish called a *diable* or "sea-devil" is a *baudroie* or *Lophius,* according to Delaunay, *Zoologie,* p. 169.

159. In the Malgaigne edition, the word is spelled "Occane," and in Céard "Oceane." It was translated by Johnson as the Ocean. Obviously, the "Ocea sea" covers a vast amount of water, in which anything might be thought to dwell

160. Olaus Magnus is Olaf Manson (1490–1557). He was archbishop o Uppsala and author of the *History of the Northern Peoples.*

161. Bergen, in Norway.

162. As Malgaigne and Céard point out, it is a question here of a seal as ca be seen by the figure. (See Pliny, *Natural History* 32, 20 and 9, 15.) At thi juncture in the early editions of *Des Monstres* came the story and figure of "Marine Boar." This passage was subsequently moved to Paré's "Discourse o the Unicorn" (q.v., Appendix 1). Early "zoos," such as the one at Fontainebleau are discussed by Callot in *La Renaissance des sciences de la vie,* pp. 50–55.

163. The Isle of Thylen is no doubt Thule or Thöl, in Greenland. Thule wa also the name given by the ancients to a northerly European island, variousl identified with Iceland, Norway, and the Shetland Islands.

164. In the editions of 1573 and 1575 the story and figure of a sea elephan came here. In 1582 Paré moved them to his "Discourse on the Unicorn" (q.v. Appendix 1).

165. Thevet calls it Mount Marzouan (Céard edition of *Des Monstres,* p. 190 n. 267).

166. Paré uses *ongle(s),* as a modern French person would, to cover severa English words, including nail, claw, hoof, talon, etc. The zoological term *unguis ungues* has been proposed in brackets as a word that might accurately reflect th sense in which Paré is using the word *ongle(s).*

167. The Johnson translation gives as the source Pliny, *Natural History* 28 8: "It is recommended as one of the most efficient remedies for cataract t anoint the eyes with crocodile's gall, incorporated with honey" (Bostock, 5:315)

168. "Insect fish" is one of the books of Rondelet's *Complete History of Fish*

169. The Johnson translation gives "sea-pricks." *Vit* is, in French, a vulga word for *penis.* (Cf. *Tire-vit.*)

170. Pliny, *Natural History* 9, 1 and 74. But, as is often the case, Paré is quoting indirectly, here through Rondelet.

171. Or, after Cotgrave and Johnson, a Eusebian, or winter pear, or a sugar pear.

172. Johnson translates this as the Sarmatian Sea, which would be located in the Lower Don region. Sarmatia is a region north of the Black Sea (Fox, p. 11). Delaunay (*Ambroise Paré, naturaliste,* p. 31) identifies this as the Baltic Sea.

173. In French a *limaçon* is a "snail" or a "conque." A *limaçon de mer,* referred to here, is a "marine snail."

174. Themistitan is Mexico City, or Temistitlan. (Atkinson, *Nouveaux horizons,* p. 55. Delaunay (*Ambroise Paré, naturaliste,* p. 37) claims the "lac doux" is not Lake Texcoco, but the nearby fresh water, well-stocked Lake Chalco. He is unable to identify the "Hoga."

175. I.e., headwaters or pipes for diverting the river in order to fish it.

176. *Pas* was a measure of two feet and a half; or three and a half; or five (depending on the locality). Hence, the fish leaps anywhere from 250 to 500 feet, since Paré writes "cent pas."

177. An example of the *discours indirect libre* since Lery, not Paré, is speaking in the first person.

178. I.e., Chioggia in the Gulf of Venice.

179. An *aulne* or ell was about 4 feet, depending on the particular locality.

180. Elian, *History of Animals* 8, 31 and Pliny, *Natural History* 9, 51. According to Cuvier, the *pinnotheres,* discussed by Paré in the next paragraph, is Bernard the Hermit, or the hermit-crab. Pierre Gilles, an ichthyologist of note, made a compilation of Elian's *De vi et animalium natura* in 1533.

181. The naker or nacre, from which comes mother-of-pearl.

182. "Humanity never ceases to like monsters, and it finds them wherever they are" (Baltrušaitis, *Réveils et prodiges,* p. 332).

183. *Fresque.* Huguet says of this word as it appears in Paré's *Des Monstres,* "Il faut lire *fusque* et non *fresque.*" *Fusque,* from the Latin *fuscus: brun, noirâtre sombre.*

184. Paul Delaunay, in *Ambroise Paré, naturaliste,* p. 38, states that Saflingen is near Doel, whereas "*Hastingue* is an unknown place in Belgium." All the geographical names in the passage refer to places and rivers in Belgium.

185. Remora, the suckstone or sea lamprey. As Paré says, Pliny discusses this famous creature (*Natural History,* 32, 1, "The Echeneïs," and 9, 25 [41]).

186. Pliny narrates this tale, which Bostock terms "an absurd tradition, no doubt, invented, probably, to palliate the disgrace of his defeat" (Pliny, *Natural History,* 32, 1; Bostock, 6:2). The naval victory of Octavian and Agrippa over Antony and Cleopatra occurred in 31 B.C. at the Greek promontory of Actium, at the entrance to the gulf of Ambracia, today called Arta.

187. This translation is taken from Du Bartas, *His Devine Weekes and Works,* translated by Joshua Sylvester (London: Mumfrey Lownes, 1605), "The Fifth Day," pp. 159–60. Du Bartas was very much admired by Paré; he used him and

his commentator Goulart heavily in some passages of *Des Monstres,* and also quoted him in chap. 28 of his book *On Venoms* (Malgaigne, 3:283–349). [Paré is referring here to the electric ray, or *Tetronarce nobiliana,* family *Narcobatidae* WBJ]

188. At the time of the great explorations (about 1492) the continent was conceived of as "Africa and Ethiopia." See Fox, *Atlas of European History,* pp 26–27. Pliny discusses the ostrich, in *Natural History,* 10, 1.

189. The Mareschal de Rets, Albert de Gondi (1522–1602), was made maréchal of France and the governor of Provence in 1573. His land (Retz) was raised by Charles IX to a *duché-pairie* in 1581. Albert de Gondi was father of Henri de Gondi (1572–1622), the celebrated Cardinal de Retz. Paré mentions the Mareschal de Retz again in chap. 31. (His wife, from whom he obtained the name Retz, was the brilliant *saloniste* sketched by Bailly, *La Vie litteraire,* pp 219–33.)

190. A span equalled nine inches.

191. One *doigt* equalled the sixteenth of a foot. In other words Paré says each bone equals nine inches.

192. I.e., as a preeminent example.

193. Zocotera, Socotra, south of the Arabian Peninsula.

194. The Huspalim is called a Hulpalis in Johnson, after Thevet, and also according to the 1579 and the 1582 (Latin) editions.

195. The information is taken from Thevet, but is not accurately quoted Moreover, one might note that Caragan is a mountain in Australia, a region known only dimly during the Renaissance.

196. One *toise* equalled a fathom, or about 6 feet.

197. The edition of 1579 gave four articles at this juncture, three of which were subsequently moved to the *Discourse on the Unicorn.* These three articles (on the *Pyrassouppi,* the *Camphurch,* and the *Bull of Florida*) will be found in Appendix 1.

198. Pliny, *Natural History,* 8, 1–12, as well as Plutarch, Suetonius, and Elian are some of the sources for Paré's developments on the elephant.

199. Leo Africanus (1465–1550), called John, was an Arab geographer. The English translation of his description of his journeys through Africa, originally written in Arabic, was reissued in three volumes by the Hakluyt Society in 1896 and a reprint was made in 1963 under the title *The History and Description of Africa.* Céard (*Des Monstres*, p. 196) claims that Paré's reference is not found in Leo Africanus, but rather in Joannes Boemus's *Omnium gentium mores* translated into French in 1540.

200. Here followed a passage on the Rhinoceros in the editions of 1573 and 1579, subsequently moved to the "Discourse on the Unicorn"; see Appendix 1).

201. Matthiole (corruption of Mattioli). Pietro Andrea Mattioli (1500–1577) was the author of *Commentaires on Dioscorides* (Venice, 1554, in fol.). Despite errors, this work is still considered a precious document in the history of medicine. His *Opera omnia* was published in Basel in 1598. Later editions appeared in 1674, 1712, 1744.

202. References in the Johnson translation are to Pliny, *Natural History,* 8, 33, and to Aristotle, *History of animals,* 2, 12.

203. Paré himself states the additional, and often indirect, sources for his observations on comets and on astronomy later on in this same chapter. They are: Ptolemy, Pliny (*Natural History,* 2, 22), Aristotle, Milichius (commentary on Pliny), Cardan (commentary on Ptolemy), and especially Du Bartas (as well as his commentator Goulart). See also Jacques de Voragine, *La Légende dorée,* 1:33.

204. The Comet of Westrie. Johnson gives "Uvestine." Boaistuau gives 1527 as the date of this comet. The literature on comets often cites Paré's description of the "comet of 1528." Some authors erroneously take it to be a first-hand account.

205. Flavius Josephus (A.D. 37–95) is the author of *Jewish Antiquities.* Eusebius (A.D. 265–340), was bishop of Caesarea and author of an *Ecclesiastical History.*

206. Agrippa d'Aubigné has no doubt given us the most celebrated description of this manifestation of hunger and fear in *Les Tragiques.* It is widespread as an example of the "monstrous" effects of famine. See, for example, the gruesome tale of a mother who devoured her child in Jacques de Voragine, *La Légende dorée,* 1:340.

207. Claudian was a Latin poet (A.D. 370–404). The quotation is from his *De bello Gothico.*

208. *Agreeable* here has the sense of "in agreement." The paraphrase of Du Bartas is taken from Goulart.

Most interesting here is Paré's disregard of Copernican heliocentric laws formulated by 1543. See on this subject George Sarton's "The Quest for Truth" in *The Renaissance,* pp. 61–64.

209. Paré should have written *de* (not *et*) as did Goulart (Céard edition of *Des Monstres,* p. 199), that is, "of opposite movements."

210. Note the strong Pascalian ring of the paragraph, which appears to be original with Paré.

211. The editor of the Club Français du Livre edition of *Des Monstres,* makes an interesting comment (pp. 366–67) on this passage: "Ce 'hola ma plume, arreste-toy', on ne saurait décider dans quelle mesure il est prudent, et dans quelle il exprime l'humilité dévote de Paré. Il ne faut pas oublier, cependant, que les considérables travaux de Copernic (*Traité des révolutions des mondes célestes*) sont connus des savants vers la même époque. Jusqu'à sa mort, Paré laissera cependant ce passage inchangé. Ce qu'il dit des planètes, à la fois chèvre et chou, n'est en rien décisif. Sans doute Paré s'en remettait-il, hors de sa spécialité, à l'opinion générale."

212. The interpreter of Du Bartas is, of course, Simon Goulart, *Commentaire* (*de la Sepmaine de Du Bartas*), 1583.

213. This is the opening of Psalm 19 in the Protestant Bible. This and the following Psalm are quoted by Paré from the celebrated poetic paraphrases of Clément Marot.

214. In Johnson, Sugolia. Lycosthenes' original locates "Suntogia" in Alsace (Céard edition of *Des Monstres,* p. 200). We see from Paré that "Sugolie" is on the border of Hungary.

215. Lycosthenes calls this "Lusatia," a region in Brandenburg, Saxony, and Silesia.

216. In Johnson, "Jubea." In Lycosthenes, Juben is referred to as an *oppidium,* a provincial town, sometimes fortified.

217. For Paré's inaccuracies in quoting Lycosthenes, see Céard's edition of *Des Monstres,* p. 200.

218. In Pliny, the name is Vectius Marcellus (*Natural History,* 7, 38). Johnson used Vectius, probably according to the Latin version.

219. In Pliny the territory is Marrucinum (*Natural History,* 7, 38). Bostock (3:527) observes that this is an "exaggerated account merely of a land-slip."

220. Citations from Pliny in this chapter are largely from book 2, chapters 57–58 and 73. See also book 17, chap. 25, for the time of Nero.

In the 1579 edition of *Des Monstres* there was a last paragraph which, in part, now appears in our chap. 39. Moreover, additional items once contained in chaps. 38 and 39 of this edition of *Des Monstres* were subsequently moved to Paré's book *On Tumors* (chap. 19). These items are reconstructed in our Appendix 2.

221. Paré quotes Ortelius here. Of course, this is not the date traditionally given for the founding of Rome by Romulus (753 B.C.), but 350 B.C. is a reasonable date for the founding of Rome as a politically important city.

222. Céard, in his edition of *Des Monstres,* p. 201, proposes that we read "in the year 254, on the first day of February. In the year 1169. . . ."

223. I.e., Tomaso Fazello, called Fazellus (Italian, 1498–1571), whose name Paré spells variously in this chapter. He is the author of *De Rebus siculis Decades duae* (1554). But did he write these "tragic stories"? Céard (p. 201) proposes Ortelius, whom Paré mentions at the beginning of this chapter, as a more likely source.

224. I.e., Agrigentum, or Agrigento, capital of the province of the same name in southern Sicily, founded in 580 B.C. by the Greeks.

225. In Du Bartas the place name is Cassaglie; in Sylvester it is Quinsay. Both refer to Hangchow in China (Marco Polo, *Voyages,* 2:64 ff.). See Holmes edition of Du Bartas, *Works,* 2:33.

226. That is, Goulart.

227. Lucio Maggio, *Del Terremoto* (1571). Paré knew Italian well, as demonstrated here and also by the Italian alternative he gives to the word *clouporte* (*porceleti*) in chap. 16.

228. Psalm 104, verse 26. Paré again cites Marot's paraphrase. The sea dragon is, of course, the *baleine,* a whale, or the Leviathan.

229. Konrad von Gesner, Swiss naturalist, 1516–65. Paré refers to his *Icones animalium.*

230. Hector Boece, or Boethius (1465?–1536?), author of *Scotorum historiae* (in 19 books), published in 1527. Boece is cited by Gesner.

231. From Thevet's *Cosmographie,* book 5, as Paré will indicate a bit later.

232. A. Dürer's portrait of the rhinoceros (1515) has been used as the basis of the crude figure reproduced by Paré (fig. 85).

233. All items in this appendix, first moved in 1582 to the *Discourse on the Unicorn,* were moved to the book *On Venoms* in 1585. Malgaigne, however, as has been noted, followed the posthumous edition, and kept these items with the treatment of the unicorn. (Brabant, *Médecins,* p. 110, incidentally, regards Paré as the "father of toxicology.")

234. *Kystis,* membrane or bladder full of fluid.

235. *Mole,* "a shapeless lump of flesh or a hard swelling in the womb" (Cotgrave); "a fleshy mass or tumor formed in the uterus by the degeneration or abortive development of the ovum" (Dorland, 25th ed.).

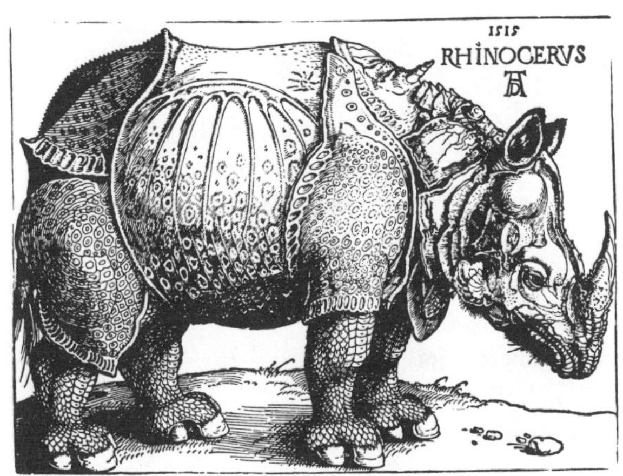

89. A. Dürer's Rhinoceros

BIBLIOGRAPHY

Askham, Anthony. *A Little Herball.* London, 1561.

Atkinson, Geoffroy. *Les Nouveaux horizons de la renaissance française.* Geneva: Slatkine, 1969.

Augustine. *The City of God.* New York: Doubleday Image, 1958.

Bailly, Auguste. *La Vie littéraire sous la Renaissance.* Paris: Tallandier, 1952.

Baltrušaitis, Jurgis. *Réveils et prodiges.* Paris: Armand Colin, 1960.

Bernard of Clairvaux. "Letter to William of St. Thierry." *A Mediaeval Garner.* Cambridge: Cambridge Press, 1967.

Béroalde de Verville, François. *Le Palais des curieux.* Paris: Chez la Veufve M. Guillemot, 1612.

Birkmayer, W., ed. *Epileptic Seizures, Behaviour, Pain.* Baltimore: University Park Press, 1976.

Boas, Marie. "The Scientific Renaissance, 1450–1630." *The Rise of Modern Science,* vol. 2. New York: Harper & Bros., 1962.

Bonham, Thomas. *The Chyrurgeons Closet.* London: Edward Brewster, 1630.

Brabant, H. *Médecins, malades et maladies de la Renaissance.* Brussels: La Renaissance du Livre, 1966.

Braunschweig, Hieronymus von. *A Most Excellent Homish Apothecarie.* Collen: Arnold Birkman, 1561.

———. *Experyence of the Worke of Surgeri.* London, 1523.

Callot, Emile. *La Renaissance des sciences de la vie.* Paris: Presses Universitaires, 1951.

Céard, Jean. *La Nature et les prodiges.* Geneva: Droz, 1977.

La Chirurgie de Maître Henri de Mondeville. Anonymous translation (1314). Paris: Firmin Didot, 1897.

Cosman, Madeleine, and Chandler, Bruce, eds. *Machaut's World: Science and Art in the Fourteenth Century.* New York: New York Academy of Sciences, 1978.

Cotgrave, Randle. *A Dictionarie of the French and English Tongues.* London: Adam Islip, 1611.

Delaunay, Paul. *Ambroise Paré naturaliste.* Laval: Imprimerie Goupil, 1926.

———. *La Zoologie au XVIe siècle.* Paris: Hermann, 1962.

Dictionnaire de l'Académie française. 2d ed. 1695. Paris: Veuve de J. B. Coignard; Geneva: Slatkine, 1968.

Doe, Janet. *Ambroise Paré: A Bibliography, 1545–1940*. Chicago: University of Chicago Press, 1937; Amsterdam: Gerard Th. van Heusden, 1976.

Dolan, John P., and Adams-Smith, Willliam N. *Health and Society: A Documentary History of Medicine*. New York: Seabury Press, 1978.

Dorland's Medical Dictionary. Philadelphia: Saunders Press, 1980.

[Du] Bartas, Guillaume de Salluste. *His Devine Weekes and Workes*. Tr. Joshua Sylvester. London: Humfrey Lownes, 1605; Gainesville, Florida: Scholars' Facsimiles and Reprints, 1965.

———. *Works*. Ed. Urban T. Holmes, Jr. 3 vols. Chapel Hill: University of North Carolina Press, 1935–40.

Du Fail, Noël. *Propos rustiques et Baliverneries*. Paris: Garnier, 1928.

Dulieu, L. *Ambroise Paré et la chirurgie au XVIe siècle*. Paris: Edition de l'Accueil, 1967.

D'Eschevannes, Carlos. *La Vie d'Ambroise Paré, père de la chirurgie*. Paris: Gallimard, 1930.

Ferguson, Wallace K., et al. *The Renaissance*. New York: Harper, 1962.

Foucault, Michel. *The Birth of the Clinic*. Tr. A. M. Sheridan Smith. New York: Pantheon Books, 1973. Originally published in France as *Naissance de la Clinique*. Paris: Presses Universitaires de France, 1963.

Fox, Edward Whiting. *Atlas of European History*. New York: Oxford University Press, 1957.

Gachat d'Artigny. *Nouveaux mémoires*. 7 vols. Paris: Debure, 1741–49.

Haag, Emile, and Haag, Eug. *La France protestante*. Vol. 8. Geneva: Slatkine Reprints, 1966.

Haliwell's Dictionary of Archaic and Provincial Words. London: Reeves & Turner, 1889.

Hamby, W. B. "Ambroise Paré." *Dictionary of Scientific Biography*, vol. 10 (1974): 315–17.

———. *Ambroise Paré, Surgeon of the Renaissance*. St. Louis: Green, 1967.

———. *The Case Reports and Autopsy Records of Ambroise Paré*. Springfield, Ill.: C. C. Thomas, 1960.

Huguet, Edmond. *Dictionnaire de la langue française du seizième siècle*. Paris: Champion, 1932.

Keynes, Geoffrey, ed. *The Apologie and Treatise of Ambroise Paré*. Tr. Thomas Johnson. Chicago: University of Chicago Press, 1952.

Le Duchat. *La Ducatania*. Amsterdam: Pierre Humbert, 1738.

Le Paulmier, Claude-Stephen. *Ambroise Paré d'après de nouveaux documents*. Paris: Charavay, 1884.

L'Estoile, Pierre de. *The Paris of Henry of Navarre*. Tr. and ed. Nancy L. Roelker. Cambridge: Harvard University Press, 1958.

Lyons, Albert S., and Petrucelli, R. Joseph. *Medicine: An Illustrated History*. New York: Harry N. Abrams, 1978.

Malgaigne, J. F. *Surgery and Ambroise Paré*. Tr. and ed. Wallace B. Hamby. Norman: University of Oklahoma Press, 1965.

Michelet, Léon. *La Vie d'Ambroise Paré*. Paris: Librairie le François, 1930.

Mondeville, Henri de. *La Chirurgie de Maître Henri de Mondeville*. Paris: Firmin Didot, 1897.

Montaigne, Michel de. *Essais*. Ed. Maurice Rat. 2 vols. Paris: Garnier, 1962.

Packard, Francis R. *Life and Times of Ambroise Paré*. New York: Paul B. Hoeber, 1921.

Paget, Stephen. *Ambroise Paré and His Times*. London: G. P. Putnam, 1899. Contains an English edition of *Journeys in Diverse Places* and fragments from Paré's treatise "On the Plague."

Pallister, Janis L. "Fifteenth Century Surgery in France: Contributions to Language and Literature." *Fifteenth Century Studies* 3 (1980): 147–51.

Paré, Ambroise. *Des Monstres, des prodiges, des voyages*. Ed. Patrice Boussel. Paris: Le Club du Libraire, 1964.

———. *Des Monstres*. Ed. Jean Céard. Geneva: Droz, 1971.

———. *Oeuvres: animaux, monstres et prodiges*. Paris: Le Club Français du Livre, 1954.

———. *Oeuvres complètes*. Ed. J. F. Malgaigne. Paris: J. B. Baillière, 1840; Geneva: Slatkine, 1970.

———. *Journeys in Diverse Places*. Tr. S. Paget. In *Scientific Papers, Physiology, Medicine, Surgery, Geology*, vol. 38. New York: P. F. Collier & Son, 1910.

———. *The Workes of That Famous Chirurgion Ambrose Parey, translated out of Latine and compared with the French* by Th[omas] Johnson. London: Th. Cotes, and R. Young, 1634; Pound Ridge, N.Y.: Milford House, 1968. Reprinted in 1649, 1665, 1678, and 1691.

———. *Three and Fifty Instruments of Chirurgery*. London, 1631; facsimile reprint, New York: Da Capo, 1969.

Pedote, V., ed. *L'Opera ostetrico-ginecologica di Ambrogio Paré*. Bologna: Cappelli, 1966. Gives résumés of the chapters of *Des Monstres*.

Petit de Juleville, L. *Histoire de la langue et de la littérature françaises* 3:511–21. Paris: Colin, 1897.

Pliny. *Natural History.* Tr. and ed. John Bostock and H. J. Riley. 6 vols. London: Henry G. Bohn, 1856.

Power, d'Arcy. "Archealogica Medica: The Iconography of Ambroise Paré." *British Medical Journal* 2 (1929): 965.

Raices, R. "Sobre la vida y obra de Ambrosio Paré." *Sociedad Argentina de Antropologia y Historia de la Medicina* 1 (1974): 4–9.

Rather, L. J. "Ambroise Paré, the Countess Margaret, Multıple Births and Hydatidiform Mole." *Bulletin of the New York Academy of Medicine* 47 (1971): 508–15.

Read, Alexander. *A Treatise of Chirurgeri.* London: Francis Constable, 1638.

The Renaissance. New York: Harper, 1962.

Rickard, Peter. *La Langue française au seizième siècle.* Cambridge: Aux Presses Universitaires, 1968.

Sadler, John. *The Sick Womans Private Looking-Glasse.* London: Stephens & Meredith, 1636.

Schneegans, Heinrich. *Geschichte der grotesken Satire.* Strassburg: Karl J. Trübner, 1894.

Scullard, H. H., and van der Heyden, A. *Shorter Atlas of the Classical World.* London: Thomas Nelson, 1962.

Stubbe, Hans. *History of Genetics from Prehistoric Times to the Rediscovery of Mendel's Laws.* Tr. T. R. W. Waters. Cambridge, Mass.: M.I.T. Press, 1972.

Temtamy, Samia, and McKusick, Victor A. *The Genetics of Hand Malformations* [Birth Defects: Original Article Series] 14, no. 3 (1978): 69.

Tilley, Arthur. *Studies in the French Renaissance.* New York: Barnes & Noble, 1922.

Vicary, M. Thomas. *A Profitable Treatise of the Anatomie of Man's Body.* London: Henry Banforde, 1577.

Vincelet, L. "Ambroise Paré et la religion." *Histoire des sciences et de la médecine* 2 (1968): 79–93.

Vigo, Joannes de. *The Most Excellent Works of Chirurgerie.* Collen: Edward Whytchurch, 1543.

Voragine, Jacques de. *La Légende dorée.* 2 vols. Paris: Garnier Flammarion, 1967.

Wickersheimer, Ernest. *La Médecine et les médecins en France à l'époque de la Renaissance*. Paris, 1905; Geneva: Slatkine Reprints, 1970.

Wright's Dictionary of Archaic and Provincial Words. London: Henry G. Bohn, 1857; Detroit: Gale Research, 1967.

Wolff, Etienne. *La Science des monstres*. Paris: Gallimard, 1948.

Zimmerman, L. M. "Humanity and Compassion in Medicine: Ambroise Paré." *Chicago Medical School Quarterly* 27 (1968): 233–34.

Zinger, Ilana. *Structures narratives du "Moyen de Parvenir" de Béroalde de Verville*. Paris: Nizet, 1979.

INDEX

Abscess, 170–71. *See also* Apostema
Actium, 134
Aeschylus, xviii
Africa, 144, 147, 149, 204n
African History (Leo Africanus), 148
Agricola, Georgius, 199n
Agrigento, 160
Ague, tertian, 63
Ahab, 86
Ahob, 142
Albania, 134
Albertus Magnus (or Albert the Scholiast), 73, 93, 199n
Albinism, 190n
Albucrasis (or Albucasis), 24, 62, 187n
Alchemy, 196n
Alleman, Henry, 53
Aloés, 119–20
Ambroise Paré and His Times, xxviii
Ambroise Paré, naturaliste (Delaunay), 177
Anatomy (Paré), 172
Ancients, 5, 38, 52, 130, 150, 182n
Andrew, 103
Androgyne(s), 26, 190n
Andromeda, 38
Andura (or Hoga), 121
Angers, xvi, 18
Anjou, 24
Antarctic, 121
Antigenes, the wife of, 37
Antiques leçons (Rhodiginus), 8, 19, 67
Anthony (Saint), 74
Antwerp, 69, 112, 134, 160
Anville (le Mareschal d'), 61
Apollonius, 102
Apologia (Apologie), xvii, xxiii, xxv, xxviii, xxx
Apostema, 49, 52, 54–55, 60–62, 192n
Aquarius, 155
Arabia, 165
Arabs, 114, 117, 165
Architecture of nature, xvi, xxvi, 5
Aries, 155
Aristotle, xxii, xxv, 3, 14, 23, 26–27, 38, 72, 115, 125, 130, 144–45, 155, 184n, 205n
Arragon, 170
Aruspices, 32
Ascot, Duke of, 89
Asia, 84
Atheists, 67, 73, 196n
Aubignac, François, Abbe d', xix
Aubigné, Agrippa d', 205n
Augustin (Captain), 53
Augustine (Saint), 3, 8, 73, 89, 92, 184n
Augustus Caesar, 23
Autun, 12
Averroes, 93
Avignon, 66

Bailey, Yolande, 187n
Baillou, 170
Balthazar, 63
Balthazar (one of the Magi), 102
Baltrušaitis, Jurgis, xxii–xxiii
Baptiste, Leon, 195n
Barbary, 79
Barque, Pierre, 59
Bassarus, Achilles, 45
Baucheron, 12
Bavaria (Duchy of), 8, 157
Bayonne, 131
Beaufort, 24
Beggars, 196n; habits of, 77, 84; in ancient times, 84; spital, 4, 74ff, 184n, 196n
Bellanger, Jean (Master), 41–42, 72
Belleforest, François, 187n
Benedict, Alexander, 48, 64
Bengal, 142
Benivenius, Antonius, xxix, 53, 59–60, 64
Bergen, 114
Bermon, 64
Bern, 24
Bernard the Hermit, 125ff

Béroalde de Verville, François, 184n
Beruce. *See* Guillaudin
Bezoar, 199n
Biarritz, 131
Bible, xxxi
Bière, Forest of, 41
"Bird of God" (or "Bird of Paradise"), 140–41
Births, multiple, 23ff, 187n
Blandy, 71
Boaistuau, Pierre, xxv, xxix, 3, 30, 66, 73, 94, 106, 150, 156, 184n
Bodin, Jean, 85, 198n, 201n
Boece (or Boethius), Hector, 163, 206n
Boiling oil, 182n
Bois-le-Roy, 41
Boistuau, Pierre. *See* Boaistuau, Pierre
Bologna, 95, 199n
Bony, Jehan, 170
Book of subtle inventions (Cardan), 106
Boscage, Monseigneur de, 63
Bouchet, Guillaume, 201n
Bourbon, the Cardinal of, 170
Bourgeois (doctor), 23
Bourges, 61
Bourg-Hersant, xvi
Bourlier, Jean, 52
Brantome, 201n
Bribon, Jacques, 42
Briot, Pierre, 90
Bristant, 16
Brittany, 61, 74
Brouet, 170
Brussels, 44, 68
Bulampech (flying fish), 121–23
Bull of Florida, 167–68
Bureau of the Poor of Paris, 79–80, 82
Burgundy, 12–13
Butrol, 167–68

Cabrolle, Barthélemy, 61
Cademoth, 165
Calabria, 160
Caligula (emperor), 135, 158
Callot, Emile, 183n
Callot, Jacques, 197n
Camota, 142
Camphurch, 166–67
Canape, Jehan, xvi
Cancellus, 125
Cancer (sign), 155
Cancer, Tropic of, 142
Canker, 74ff
Cappel, 170
Capricornus, 155
Caragan, 142
Cardan, Jerome (Hieronymus Cardanus), 67, 106, 109, 140, 155, 195n, 205n
Cassianus, 94
Castre, 110
Catania, 158–60
Causes, xxvi–xxvii
Céard, Jean, xxii, xxvi–xxvii, 163
Celée, Louis, 71
Celestial bodies, 152–55
Chambellant, 61
Chambellay, 24
Chambenoist, 71
Chameleon, 149–50
Chamoeleopardalis, 143
Champagne, 31
Champigny, 81, 82
Chancre (Sea crab), 129
Chaperon, 186n
Charles V (emperor), 157
Charles IX (king), xviii–xix, 13, 88, 96, 114, 128, 136, 139, 141, 188n, 204n
Chartier, Tiennette, 63
Chartres, 21
Chasenay, 48
Châtelet (prison), 63
Chauliac, Guy de, xxiv
Chaumière, Macé, 24
Cheval, Germain, 83
Chioggia. *See* Quioze
Chirurgie françoise. See *French Surgery*
Chronicles (Forestus), 110
Chroniques (Monstrelet), 63
Chrysippus, 55
Cimbrian Wars, 156
Cinq livres de l'imposture et tromperie des diables (Wier), 199n

Circe, 101
Ciret, René, 15
City of God (Augustine), 8, 89, 92
Civil wars, xviii
Claudian, 152, 205n
Cocquin, Pierre, 49
Coligny, Gaspard de, xviii
Collected Works (of Paré), xx; editions discussed xxiii–xxiv. See also *Oeuvres complètes.*
Collo, Jean and Laurens (sons of Laurens), 49–50, 52, 192n
Comets, 150ff
Commentaries on Dioscorides (Mattioli), 203n
Constance, 94
Constantinople, 157
Copernicus, 205n
Cornax, Matthias, 44
Corruption, teratogenic, 66ff
Cosmography (Thevet), 115–16, 119, 121, 139, 144, 146, 166–67
Courtin, Germain, 63
Cratain (shepherd), 67
Creator, the, 155
Crinay, Simon, 48
Crocodile, 115–18
Crocodilea, 116
Cromerus, Martinus, 24

Daimons (Ronsard), 196
Dalam, Jacques, 48
Dalechamps, Jacques, 24, 52, 62
Damascene, 38
Dauphiné, 48
David, 97, 156, 161
Davos, 90
De abditis rerum causis (Fernel), 97
De bello Gothico (Claudian), 205n
De conceptu et generatione hominis (Rueff), 92, 184n
De Generatione animali (Aristotle), 14, 26
de la Croix, 193n
De l'Anatomie (Paré), 188n
Delaunay, Paul, 177ff, 191n, 203n
Del Terremoto (Maggio), 206n
Demarchus, 101
Démonomanie des Sorciers (Bodin), 197
Demons, 85ff; in mines, 89ff; their deceitful ways, 91ff
De naturae divinis characterismis (Gemma), 59
De natura pueri (Hippocrates), 26
Denmark, 121
de-Pleurs, 62–63
De Rebus siculis Decades duae (Fazello), 206n
De re metallica (Agricola), 199n
D'Eschevannes, xix
Des Monstres et prodiges (Paré), xv–xviii, xxii, xxv–xxx, xxvii, xxviii, xxix, xxx, 177, and esp. 184n, 190n
Des Opérations (Paré), 192n
Des Pierres (Paré), 51. See also *On Stones*
De subtilitate rerum (Cardan), 106, 140
De tumoribus praeter naturam (Ingrassias), 171
Devils, 85ff
Dickpert, Joest, 68
Didacticism, xxv–xxvii, xxix, 155, 187n
Diet, 193n
Diphanes, 55
Diphilus, 42
Disasters, natural, 157–61
Discours de la Licorne (Paré), 163
Discourse on earthquakes, or, *Del Terremoto* (Maggio), 161
Discourse on the Unicorn (Paré), 204n, 207n
Diseases, 46ff
Diverse Journeys (of Paré), xxx
Dodens (Rembert), 195n
Dole, Bishopric of, 157
Dorothea, 24
Du Bartas, Guillaume de Saluste, xxv, 135, 154–55, 160, 203n, 205n
Dubois. *See* Sylvius
Dürer, A., 207n
Duret, 56
Dwarfism, thanatophoric, 190n
Dyseris, 52
Dysplasia, 191n

Earthquakes, 157ff
Eastern Germanic Sea, 120
Eberbach, 45
Echeneis. See *Remora*
Egypt, 87, 108
Egyptian ox, 169
Egyptians, 117
Electric ray, 203n
Elephant, 144–46, 169
Elian, 125
Empedocles, 3, 23, 38, 42, 184n
Epicureans, 89
Empiricals, 104, 200n
Endor, 97
England, 157
English Gentleman, 112
Ephesians, Book of, 73
Esdras, 3, 5, 184n
Ethiopia, 38, 144, 166, 204n
Ethiopian marmot, 141–42
Ethnics, 101
Etna, 157–58
Eusebius, 152, 205n
Experiment, Paré and, xxvi, 182n
Ezechias, 97

Facelle (or Facellus), Thomas, 158, 160, 206n
Fernel, Jean, 55, 61–62, 97, 172
Ferrara, 19, 157
Ferrara, Duchess of, 50
Flaubert, xxx
Flecelle (or Flesselles), Philippe de, 81
Floods, 160–61
Florence, 53, 59–60
Florentin, Nicole, 57
Florentines and Pisans (war between), 6
Flushing, 134
Flying fish, 121–25
Fontainebleau, 41, 114, 202n
"Foreign body," removed, 192n
Foreign matter, rejection of, 60ff
Forestus, Philippe, 110, 202n
Franc(s)-archer(s), 63–64, 195n
La France protestante (Haag and Haag), xix
Francis I, 21
Francisque, 48
French. *See* Language; *see also* Paré, his language
French Surgery (Dalechamps), 24, 52, 62

Gabelle, 195n
Gachat d'Artigny, 199n
Galen, xvi, 3, 103, 172, 184n
Gangapur, 142
Ganges River, 142
Garnier, Germain (Marie), 31–32
Gascony, 36
Gaspar (one of the Magi), 102
Gemini, 155
Gemma, Cornelius, 59
Germain, Marie, 31–32
Germain, Pierre, 17
Germany, 20–21, 89
Gesner, Konrad von (or Gesnerus, Conradus), xxix, 109, 112, 128, 163, 206n
Gilles, Pierre, 203n
Giraffe, 142–44
Girande, Cypriane, 21
Gislinger, John (doctor), 24
God, 3–7, 73, 85–87, 90, 95, 102–3, 105, 125, 152, 154–55, 160
Good Friday, 74
Gourdon, 57
Goulart, 204n, 205n
Gourmelen, Estienne, xvii, xx, xxiii–xxiv
Grangier, xxv
Grasbonnet, 59
Greek, 143
Grillant, Paul (or Grillando, Paolo), 88, 199n
Grisons, Canton of, 90
Gros, 63
Guelders, 37
Guerrier, Guillaume, 49
Guillaudin (or, Wieland)-Beruce, Melchior, 41
Guillemeau, Jacques, xx, xxviii, 55, 61
Guillement, François, 61
Guinter, Johann, xxi
Guion, 61

Guise, 33
Gunshot treatment, 183n

Haag and Haag, xix
Haiit, 147–48
Hair, ingrown, 64, 195n
Hamby, xx
Harelip, 191n
Harmois, Court of, 41
Hastings-au-Doel, 134
Haultin (or Hautin), Jean, xx, 36
Headache, 55
Heidelberg, 30
Heliodorus, 38
Henri III, xix
Hermaphrodites, 24, 26ff, 188n; laws governing, 27; in Aristotle's *Problems,* 27
Hermit (Crab), Bernard the, 125ff, 203n
Hertages, Egidius, 44
Hidustes, King, 38
Hippocrates, xxx, 3, 8, 26, 37–40, 42, 52, 65, 84, 184n
"Histoire digne," 170
Histoires prodigieuses (Boaistuau), 3, 66, 106, 150, 156, 184n
Histoires prodigieuses (Lycosthenes), 58
Historia naturalis (Pliny), 184n
Historiographers (Greek), 143
History of Africa (Leo Africanus), 148, 189n, 204n
History of Animals (Aristotle), 125, 145
History of Animals (Gesner), 128
History of Brazil (Léry), 118, 122
History of Ethiopia (Heliodorus), 38
History of Fish (Rondelet), 125
History of the Northern Peoples (Olaus Magnus), 200
History of Poland (Cromerus), 24
Hoga (or Andura), 121
Holland, 161
Hollier, 83
Hollier, Jacques, 52, 83
Holy Scripture, 86, 88, 97, 99, 100, 104
Homer, 101
Homo signorum, 196n
Hôtel-Dieu, xvii, 63, 195n
Houlier, Jacques, 55
Huguet, 61
Humanism in medicine, 182n
Hunchbacks, 55
Hungary, 156
Huspalim, 141–42

Illness from supernatural causes, 97ff
Illusions, diabolical, 94ff
Illustrations, discussion of, xxviii–xix, 5, 183n
Illyrian Sea, 112
Imagination in parents, 8–9, 38ff, 54–55
Impregnation, diabolical, 92–94
Incubi, incubus, 91, 105
India, 31, 142, 144
Indians, 146
Indic language, 140
Ingrassias, Jean Philippe, 171–72
Injury, prenatal, 47ff
Insect fish, 118
Introduction (Paré), 192n, 193n
Italy, 7, 9, 18, 24, 30, 95, 146, 157
Ivory, 146

Jackson, William B., xxxii, 177
Jacob, 38
Jacques, 64
Javelle, Joseph, 50
Jean (a joiner), 59
Jeanne (changed to Jean), 31
Jehu, 89
Jeremiah, 73
Jerusalem, 79, 152
Jesus Christ, 4, 85–86, 94; Passion of, 152
Jewish Antiquities (Josephus), 205n
Jews, 152
Jezebel (Queen), 89
Job, 86, 97
John the Evangelist (Saint), 4
John the Baptist (Saint), 38
Johnson, Thomas (translator), xxix, xxvii–xxix
Joint mouse, 192n

Josephus, 152, 205n
Joubert, Laurent, 57, 61, 193n
Journeys in Diverse Places (of Paré), xxviii
Juben, 156
Judaic nations, 152
Judas, 86
Julius II (pope), 7
Julius II and Louis XII, war between, 7
Jupiter, 152, 154

Karenty, 97
King of the Stars, 153ff
"Knotting the point," 98, 105ff, 201n
Krakow, 24, 58

Laban, 38
La Guillautière, 161
Lamie, 127–29
Langius, Joannes, 45, 96
Language, of Paré and the Renaissance, xxix–xxx
Languedoc, 125
Lansac, Lord of, 96
La Pallude, Pierre de, 92
Lardot, Etienne, 42
Larissa, 52
La-Roche-sur-Yon, Prince of, 131
Laval (in Maine), xvi
Lavater, Louis, 89, 90, 199n
Lectionem Antiquarum (Rhodiginus), 184n
Le Fèvre (or Lefèvre), Pierre, 63
Le Grand, Claude, 59, 192n
Le Grand, Nicole, 52–53
Le Loyer, 200n
Leo Africanus, John, 148, 189n, 204n
Leo (the sign), 155
Lenoncort, Cardinal of, 32
Leon, Baptiste, 66
Leoniceno, Nicolao, xxi
Lepers, 198n
Leprosy, 76ff, 121
Lery, Jean de, 118, 122
L'Estoile, Pierre de, xx
Letters on Medicine (Langius), 96
Levinus, 58
L'Heister, 89
Leviticus, 5
Libra, 155
Liège, 68
The Life and Times of Ambroise Paré (by Packard), xxviii
Limp mimicry, 191n
Linacre, Thomas, xxi
Lisbon, 157
Livre de la Generation (Paré), 187n
Lombardy; "Lombard brothers," 57
Loqueneux, Antoine, 31
Lorraine, 22
Loretto, Our Lady of, 79
Louis XII, 7
Louis (emperor), 157
Louvain, 59
Lusalia, 156
Lusitanus, Amatus, 31, 189n
Luther, 191n
Lycosthenes, Theobald Wolfhart (Conradus), xxv, xxix, 3, 8, 58, 150, 157, 184n
Lyons and Petrucelli, xv
Lyons, Archbishop of, xx

Macerie, 109
Macrocosm/Microcosm. *See* Microcosm macrocosm
Magdaleine, 94
Magdeburg, 157
Maggio, Luccio, 161, 206n
Magic, 95ff. *See also* Remedies, magic
Magicians, kinds of, 95
Magnus, Claus, xxv
Maine, 24
Maldemeure, Sire de, 24
Malgaigne, xv, xxiii–xxiv, xxvi–xxvii xxix–xxx, xxxii
Malingerers. *See* Beggars
Malta, Isle of, 159
Mansfeld (or Mansfelt), Charles, Coun of, 56, 193n
Manucodiata, 140–41
Marcel, Bishop, 110
Marcellus, Vasseus, 157, 206n
Marchant, Jacques, 21

Marconville, Jean de, 95
Marco Polo, 160
Marcus Antonius, 135
Marescot, Michel, 63, 170
Marguerite, 24
Marine Boar, 163
Marine Lion, 109–10
Marine Sow, 114–15
Marius, 156
Marly, 51
Marot, Clément, 205n
Marquette, 33
Marrucin, 158
Mars, 152, 154
Martin d'Arles, 92, 199n
Martin V, Pope, 66
Martin, Gaspard, xvi
Marvels, defined, 3
Matthiole (or Mattioli), 150, 204n
Mauretania, 144, 189n
Maximilian, King of the Romans, 44
Mazelin, Jehanne, xvii
Mazille (monsieur de), 96
Mazovan, Mount, 114
Medical Observations and Rare Examples (Valescus de Tarante), 63
Melchior (one of the Magi), 102
Melchior (Wieland). *See* Guillaudin
Melted lead, hands washed in, 106
Melun, 42, 71
Mena, 108
Menstruation, 5–7
Mercury, 152
Mermaids, 107
Messina, 160
Metz, 13, 22
Meudon, xviii, 66
Mexía, Pedro, 191n
Mexico, Kingdom of, 121
Michel, 69
Microcosm/macrocosm, xvi, 53ff, 60, 192n
Midi, 152
Milan, 106
Milichius, 155, 205n
Milot (monsieur), 63, 170–71
Moderation, xxii, 105, 182n
Mole, 61–62, 171, 173, 207n
Mollin, Jean, 69
Molucca(s), 140, 166; king of, 167
"Monk-fish," 108
Monoceros, 169
Monsters: causes and definition of, xxvi, 3; celestial, 150–57; derivation of word, 196n; laws governing, 188n; meaning extended, 130; and morality, 188n, 195n; moral significance, 186n; terrestrial, 157–59; treatises on, as a genre, xvi, xxii–xxiii
Monstrelet, Enguerrand de, 63
Monstrous animal from Africa, 148–49
Montaigne, xv, xxii, xxxi, 187n, 189n, 190n, 196n, 199–200n
Montargis, 50
Montelimar, 48
Montmartre, 84
Montmorency, Maréchal of, 96
Montpellier, 61
Moon, 152ff
Moor, Moors, 38–39, 142, 147, 171–72, 191n
Morphology. *See* Structure
Moses, 5, 38, 95
Mount Etna, 157–58
Multiple births, 23ff, 187n
Multiplicity, xv
Munster, Sebastian, 16, 187n
Muret, Captain, 48
Muteness, feigned, 79–80, 84

Naples, 57, 146
Natural disasters, 157–61
Natural History (Pliny), 107, 129, 184n
Nature, xxvi, xxvii, 3, 5–9, 33, 38, 60–61, 67, 71, 73, 94, 104–7, 125, 130, 145, 163, 172
La Nature et les prodiges (Céard), xxii
Naticus, Natilus, 129–30
Nebuchadnezzar, King, 101
Nenzesser, Ulrich, 96
Nero, 101, 157
Netherlands, 141

Nile, 108, 116
Noël du Fail, xxx, 197n
Normandy, 80, 83
Nymphes, 188n

Obsequens, Julius, 157
Observation (in Paré), xii
Observations (Dodoens), 195n
Ocean Sea, 112, 163
Oeuvres complètes (of Paré), xxiii, xxv, xxvii, 163
Olaus Magnus (Olaf Manson), 112, 114, 202n
On Air and Waters (Hippocrates), 84
On Epidemics (Hippocrates), 37, 52, 65
On Errors Made by the People (Joubert), 57
On Fish (Rondelet), 108, 127
On Handgun Wounds (Paré), 193n
On Monsters. See *Des Monstres et prodiges*
On the Plague (Paré), xxviii, xxxi, 60, 193n
On Reproduction (Hippocrates), 42
On Simples (Galen), 103
On Stones (Paré), 191n
On Surgical Operations (Paré), 193n
On Subtlety (Cardan), 140
On Tumors in General (Paré), 170, 193n, 206n
On the Unicorn (Paré), xxvii, 163, 204n, 207n
On the Variety of Things (Cardan), 67
On Worms (Paré), 192n
On Venoms (Paré), 204n, 207n
"Oriental Countries" (book on, by Marco Polo), 160
Orleans, 64–65
Orobon, 114–15
Ortelius, Abraham, 158
Osteochondritis dessicans, 192n
Osteogenesis imperfecta, 190n
Ostrich, 136–39
Otto III (emperor), 157

Pacheca, Marie (or Manuel), 31
Packard, xxiv–xxv, xxviii
Paget, Stephen, xviii–xix, xxi–xxii, xxviii
Palatinate, 30, 188n
Palermo, 160
Palissy, Bernard, 184n
Pallister, Janis L., 183n
Pallister, Philip D., xxxii, 177
Pallude, 199n
Palpitation of the Heart (Hollier), 52
Pamphile, 103
Pape, Jacques, 48
Paracelsus, 192n
Parasitic snake, 83ff
Paré, Ambroise: and the Ancients, xxi–xxii; and the army, xvii–xix, 181–82n; birth of, xvi; borrowings of, xxv–xxvi, xxxi–xxxii; as a businessman, xviii; and classicism, xxii, 182n; his contributions to medicine and surgery, xxi; death of, xx–xxi; his education, xvi; and his family, xvi–xx; and the French language, xx, xxiv; as a humanist, xv, xxi–xxii; use of illustrations, xvi, xxix; his language, xxix, 184n, 191n; and the medical establishment, xix–xx, xxiii–xxv; as moralist, 196n; and observation, xvii, 47; his religion, xv–xvi, xvii–xix, 182n, 187n; as storyteller, 196n; his style, xvi, xxviii, xxx–xxxi, 195n; as a teacher, xxv, xxvii, xxix, 155 (*see also* Didacticism); and toxicology, 206n
Paris, xvi–xviii, xx, 10, 13, 17, 21, 33, 36–37, 49–50, 61, 63, 72, 79, 81–82 146, 197n
Paris, the executioner of, 103
Parlan, Catherine, 49
Parpeville, 33
Parrot, 115
Pascal, 205n
Pasquier, E., 201n
Paul III, Pope, 110
Paul (Saint), 3, 86, 184n
Pausanius, 168
Peloponnesus, 24
Pernelle, Matthée, 17
Persina, Queen of Ethiopia, 38
Petit, Esme, 42

Pharaoh, 101
Philonium (Valesco de Taranto), 195n
Pica, 199n
Pico della Mirandola (or Picus Mirandula, Franciscus), 24, 187n
Piedmont, 10
Piedmont, Prince of, 12
Pigray, Pierre, 79
Pilomatrixoma, 194
Pinna, Pinne, 125–27
Pineau (Master Surgeon), 170
Pinothera, 125
Pisces, 155
Plate River, 165
Plimatic, 142
Pliny, xxv, xxxi, 3, 24, 32, 55, 101–2, 107, 125, 129, 134–35, 155, 157, 169, 184n
Plutarch, 125
"Point-knotters," 98, 105ff
Poland, 109
Pollux, Julius, 172
Pompey, 169
Pontanus, Jovianus, 20
Pont de Cé, 18
Popular Errors concerning Medicine and Health Regimens (Joubert), 193n
Portugal, 31, 157
Possession (demonic), 88ff, 99–101
Poulet, Jean, 71
Pratique (Benedict), 64
Pratique (Houlier), 55
Presle, Collège de, 61
Printemps (Yver), 198n
Privilège, 72
Problems (Aristotle), 14, 27, 72, 184n
Prodiges (Lycosthenes), 58, 184n
Prodigiorum ac ostentorum Chronicon (Lycosthenes), 184n
Prodigious Stories (Boaistuau), 106. See also *Histoires prodigieuses*
Psalms, 156, 161
Psychic consciousness (Paré's recognition of), 201n
Ptolemy, 154–55, 205n
"Pyrassoupi" (species of Arabian unicorn), 165–66
Pythoness, 101

Quarrel of the Ancients and the Moderns, xxii
Quiers (Chieri), 10, 186n
Quinsay, 160, 206n
Quioze (Chioggia), 123, 125, 203n

Rabelais, xxx, 183n, 198n
Rare, Paré's interest in the, xxvi, xxvii, 136
Rather, L. J., 187n
Ravenna, 7, 123, 184n
Rebours, 170
Red Sea, 114, 146, 165
Reims, 31
Remedies, magic, 101–4
Remora, 134–36
Renard, Pierre, 33
Reproduction (linked to monsters), xxiii
Republic (Bodin), 85, 198n
Rets (or Retz), Le Mareschal de, 96, 136, 139, 204n
Réveils et prodiges (Baltrušaitis), xxii
Rhinoceros, 168–69
Rhodiginus, Caelius (also Coelius), 8, 19, 67, 184n
Rhone, 161
Riolan, 170
Rita Christina, 184n
Robarchie, 30
Rohan, 61
Rolant, Isabeau, 170–71
Roman officer, 158
Romans, 152
Rome, 88, 12, 146, 158, 161, 169
Rondelet, 109, 118, 125, 127–28, 131, 193n
Ronsard, Pierre de, 182n, 198n
Rouen, 80
Rousselet, Claude, xix
Rousselet, François, xix
Rousselet, Jacqueline, xix
Rousset, François, 50
Rueff, Jacobus (or Jacques), xxix, 31, 92, 184n

Saflinghe, 134
Sagittarius, 155
Saint André des Arts (church), xvii–xxi, 33
Saint-Aubam-aux-Baronniers, 48
Saint-Bartholomew's Day Massacre, xviii
Saint Benedict Prisons, 79
Saint Claude, 79
Saint-Côme, Confraternity of, xviii
Saint-Didier, 64
Saint Eustache, 53
Saint Fiacre's disease, 77, 81, 198n
Saint Germain Faubourg, xviii
Saint Germain des Prés Hospital, 80
Saint Gervais, 170
Saint-Hilaire, Geoffroy, 184n
Saint Hubert, 79
Saint Jean d'Angely, 48
Saint John's disease, 77, 84, 198n
Saint Leo (church of), 159
Saint Main, 79
Saint Main's disease, 77, 84, 198n
Saint Mathurin, 79
Saint-Maure-des-[les-]Fossés, 51, 63
Saint Michael, Order of, 100
Saint Nicholas des Champs, 17
Saint Pris, 96
Saint Quentin, 31
Sainte-Beuve, xv
Saintonge, 127
Salt, tax on, 71, 196n
Sarboucat, Magdaleine, 42
Sarmatian Sea, 120
Sarret (story about), 193n
Sarte, 24
Satan, 88–89, 93, 96
Saturn, 152, 154
Saul, King, 97
Savelli, Bonaventure, 24
Saxony, 41, 156–57
Scheldt River, 134
Scorpius, 155
Scotland, 163
Scotorum historiae (Boethius), 206n
"Sea-Devil," 112
"Sea Elephant," 163
"Sea Horse," 112
Sea Panaches, 118
Seaux, 24
Seed: abundance of, 8ff; apt for reproduction, 92–93; lack of, 33ff; mixing of, 67ff
Senecy, Baron de, 13
Sens, 43
Sepmaine, La (Du Bartas), 135, 155, 160
Serenus, 103
Sex (changes in), 31ff
Sezanne, 71
Sicily, 158–60, 171
Siena, 24
Signs, monsters as, xxvi
Silva de varia lección (Mexía), 191n
Simon (the Magician), 101
Sirens, 107
Sky, 152ff
Snail, 120
Snake (feigned), 83ff
Snakes (as parasites), 58, 66
Sodomists, sodomites, 67, 73, 196n
Sol, 152
Sommières, 61
Sorcerers, 85ff
Sources, Paré's use of, xxi–xxxii, 184n. *See also* Paré, borrowings of
Spaniards, 121
Spanish Isle, 119
Speech (defects), 191n
Spital Beggars. *See* Beggars
Stammering, 47
Stecquer, 41, 191n
Stoics, 23
Stones, 47, 49ff, 63, 192n
Storytelling (in Paré), 196n
Structure, xvi, xxvi, 5
Stuttering, 47, 191n
Style indirect libre, xxx, 203n
Style of writing, xxx–xxxi. *See also* Paré his style
Succubi, 91
Sugolie, 156
Sun, 153ff
Supernatural, xv

Superstition, 102ff
Supplementum chronicorum orbis (Forestus), 201n
Surgical operations (Paré), 189n
Switzerland, 24
Sybaris, 67
Sylvius, xvii
Synopsis chirurgiae (Gourmelen), xxiv
Syracuse, 160

Tabourot, 200n
Tarante, Valescus de, 63, 195n
Taurus, 155
Taxonomy, xxvi
Teeth, 64
"Temple," 197n
Tendons, 65
Tesserant, Claude, 3, 184n
Tessier, Estienne, 64
Thanacth, 146–47
Théâtre de l'univers (Ortelius), 158
Themistitan, 121
Thevet, André, xxv, xxvi, 115–16, 119, 121, 139, 141, 144, 146, 166–67
Thylen, Isle of, 114
Tiber, 161
Tiesserand, Claude, 3
"Tire-vit," 51
Toads, spontaneous generation of, 66
Toucan, 139–40
Touraine, 16
Tours, 15
Tractatus de hereticis et sortilegis (Grillant), 198n
"Tragic stories" (Fazellus), 160, 206n
Tropic of Cancer, 142
Translations of Paré's works, xxvii–xxviii, 183n
Treatise on Superstition (Martin d'Arles), 199n
Treatise on the Suppression of Urine (Paré), 193n
Trichobezoar, 199n
Tritons, 107ff
Trois livres des apparitions des esprits (Lavater), 199n
Tuguestag, 96
Tumors, 170ff
Turin, 10
Tyrrhenian Sea, 110

Ulysses, 101
Unicorn, 163, 165, 167–68
Uzès, Duke d', xxiii

Varades, 170
Variety, xxvi–xxvii, 125
Vasseus Marcellus, 157, 206n
Vellay, 64
Venetians and Genevans, war between, 30
Venetians, Sea of, 123
Venice, 63, 121, 123, 146, 157, 160
Venus, 152
Venus, Madame, 98
Verdun, 59
Verignel, Charles, 64
Verona, 6, 9, 71
Viabon, 21
Vialot, xvi
Viard, Claude, 53, 79
Victoire, 64
Vigo, Johannes de, xv
Ville-du-Bois, xviii
Ville-franche-de-Beyran, 36
Violaine, 53
Virboslaus, Count, 24
Vircy, Jean de, 42
Virgo, 155
Vit-volant, 118
Vitré (in Brittany), xvi, 74
Vitry-le-François, 31
Volateranus (or Volterrannus), Raphael, 67, 195n
Volcanos, 158–60
Voragine, Jacques de, 188n, 195n, 205n
Von Braunschweig, Hieronymous, xv
Voyages faits en divers lieux, xxviii
Voyages Made into Divers Places, xxviii

Wechel, André, xxiii
Western Germans, 142

Westrie, 151
Whale, 130–34
Wier, John, 199n
Wittenberg, 156
Woman, imperfection of, 32–33, 189n
Womb: foreign matter in, 56–57; smallness of, 14–15, 42ff; suffocation of, 54, 62
The Workes of that Famous Chirurgion Ambrose Parey, xxviii
Worms (city), 16
Worms (as parasites), 55, 58–60, 63

Yver, Jacques, 197n

Zeeland, 161
Zocotera, Island of, 141
Zodiac, 152, 154, 196n
La Zoologie au XVIe *Siècle* (Delaunay) 177
Zurich, 31, 90